Challenges for African Agriculture

Challenges for African Agriculture

Jean-Claude Devèze

Editor

Preface by Jean-Michel Debrat

A copublication of the Agence Française de Développement and the World Bank

1818 H Street NW
Washington DC 20433
Telephone: 202-473-1000
Internet: www.worldbank.org

1 2 3 4 13 12 11 10

ISBN: 978-0-8213-8481-7
eISBN: 978-0-8213-8515-9
DOI: 10.1596/978-0-8213-8481-7

Library of Congress Cataloging-in-Publication data

Challenges for African agriculture / Jean-Claude Devèze, editor.
p. cm. (Africa development forum series)
Includes bibliographical references and index.
ISBN 978-0-8213-8481-7—ISBN 978-0-8213-8515-9 (electronic)
1. Agriculture—Economic aspects—Africa, Sub-Saharan. 2. Agriculture and state—Africa, Sub-Saharan. 3. Agriculture—Social aspects—Africa, Sub-Saharan. I. Devèze, Jean-Claude. II. Series: Africa development forum.
HD2117.C465 2010 2011
338.10967—dc22

2010033801

Cover photo: Arne Hoel/The World Bank
Cover design: Naylor Design

This book was originally published in French [*Défis agricoles africains,* ISBN 978-2-8111-0011-7] by Editions Karthala (Paris, 2008). The English translation was done by the World Bank General Services Department's Translation and Interpretation unit.

Africa Development Forum Series

The **Africa Development Forum** series was created in 2009 to focus on issues of significant relevance to Sub-Saharan Africa's social and economic development. Its aim is both to record the state of the art on a specific topic and to contribute to ongoing local, regional, and global policy debates. It is designed specifically to provide practitioners, scholars, and students with the most up-to-date research results while highlighting the promise, challenges, and opportunities that exist on the continent.

The series is sponsored by the Agence Française de Développement and the World Bank. The manuscripts chosen for publication represent the highest quality in each institution's research and activity output and have been selected for their relevance to the development agenda. Working together with a shared sense of mission and interdisciplinary purpose, the two institutions are committed to a common search for new insights and new ways of analyzing the development realities of the Sub-Saharan Africa Region.

Advisory Committee Members

Sub-Saharan Africa

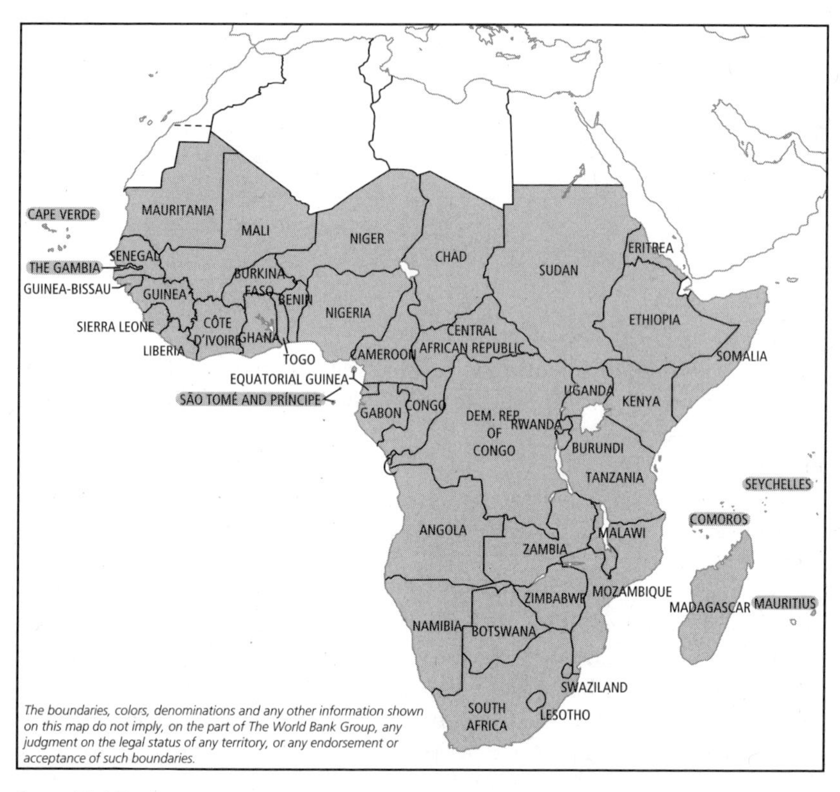

Source: World Bank.
Note: African countries not shaded above are part of the region of Middle East and North Africa as defined by the World Bank.

Contents

Boxes

Figures

Map

Tables

Preface

A geographer at the outset of my career, I have always been fascinated by the different ways in which agriculture may evolve. So, when Jean-Claude Devèze proposed the idea, some five years ago, of focusing on the future of Sub-Saharan family farms, I immediately supported his project and subsequently encouraged him to continue this work after his retirement in order to publish a collective work on this fundamental subject. Here is the result.

Naysayers predict that Sub-Saharan Africa will face ever greater difficulty in producing all the food that it needs and will therefore have to import more and more. Is there an inescapable vicious circle in rural Africa that will inevitably lead it down the path to dependency, to underemployment and rural exodus, to desertification and thus to famine, with conflicts only growing worse as a consequence? We believe that it is quite possible to initiate and support the more virtuous circles of human capacity building, economic capitalization, and improvement of natural capital.

The statistics underscore the importance of this debate. The rural population of Africa is already more than 500 million, with 80 percent living in poverty, and there can be no doubt that this population is going to grow rapidly and densities are going to increase, notwithstanding the effects of rural exodus.

There are many potential "disaster" scenarios, which could be brought on by a constellation of factors: nonownership of technical progress, a lack of credible agricultural policies, a continuation of insufficient public investment, degradation of soils and ecological capital, ineffective population policies and a disorderly rural exodus, or any worsening of already fragile situations of civil peace.

While this description is not totally implausible, it fails to give credence to a number of highly positive elements in the recent evolution of rural Africa. First, the family farms of today have a new face. Professional organizations of farmers have developed viable structures and won their autonomy in relation to national agencies and donors alike. Second, multiple actors in the "value chain" extending from the field to the consumption of food products have proved capable of coming up with endogenous innovations that are less risky and well suited to local markets. Third, technical innovations that are now being tested and disseminated provide hope that cropping systems can successfully reconcile intensification and sustainability.

Finally, African population growth and the instability of world prices for agricultural products suggest that the primary market for African farmers lies in food products headed to regional markets. It is time not for despair, but for a call to action. Everything that relates to food for people of course raises legitimate emotions, but the proliferation of emergency measures should not be allowed to overshadow the substantive measures that are necessary if African agriculture is to adapt to the new structural realities of the marketplace.

Starting at the local—if possible, decentralized—level, and moving up toward regional organizations, it is now conceivable to develop a dual path of social structuring and growth of value added at each stage of an agricultural industry. Based on past experience, it is also possible to think about financing rural development by tying it to commercial functions and investment, promotion, and dissemination policies on technical innovation. Moreover, appropriate land policies can not only provide farmers with security, but also encourage gains in productivity and sustainable management of the natural capital. In this context of stakeholder mobilization, government policies encouraging a commitment by African officials to work with their constituents on a continuing basis should carry greater weight.

This book is a collective effort by Agence Française de Développement collaborators, their partners in research units and associations, and African officials who have offered their vision of the future and what the priorities should be. I hope that it will contribute not only to changing the image of the African farmer held by African officials and their partners, but also to mobilizing the human potential needed to meet the challenges for agriculture over the long haul.

Jean-Michel Debrat

Contributors

Philippe Chedanne
Veterinarian and agroeconomist, successively technical expert in the rural development sector in Africa (Chad, Gabon, Guinea, Mali) and then responsible for rural development, agricultural policy, and food security issues at the Ministry of Foreign Affairs. Currently heads the External Relations Division at AFD after working on "trade and development" issues there.

Jean-Claude Devèze
Agroeconomist, secretary of Inter-réseaux Développement rural (a rural development networking and discussion forum), editor of the publication *Grain de sel*. He has been responsible for "rural development" projects at AFD for 30 years and has led working groups on activities in the rural sector, on the future of cotton-growing areas, and on the humid tropics. He is the author of *Le réveil des campagnes africaines* (Editions Karthala 1996).

François Doligez
Agroeconomist. Program manager at IRAM (Institut de Recherche et d'Applications des Méthodes de développement), an independent organization specializing in applied research in development methodology (www.iram-fr.org), and professor and researcher in the Economics Faculty at the University of Rennes 1.

N'Diogou Fall
Senegalese farmer, former Chairman of the West African Network of Farmer and Producer Organizations (ROPPA), and former President of the Federation of Senegalese Nongovernmental Organizations (FONGS).

Guy Faure
Researcher at CIRAD, Doctor of Economics specializing in understanding innovation processes, farm management, farm advisory services, and strengthening of producer organizations, with many years of experience in West Africa

and Central America. He is the author of *Le conseil à l'exploitation familiale* (GRET-CIRAD 2004).

Francis Gendreau
Engineer, statistician-demographer, specialist in population issues in the developing world, particularly Africa. He has worked in Madagascar, Senegal, and Cameroon and participated in missions in most African countries. He has numerous publications to his credit, including *Démographies africaines* (Estem 1996).

Jean-Jacques Goussard
Ecologist and head of mission in the firm EOS.D2C. He is involved in various fields: development of tools and reference works for assessing public policy on sustainable development, territorial management of biodiversity, adaptation to climate change, and integrated management of coastal areas.

Philip M. Kiriro
Kenyan farmer, President of the Eastern Africa Farmers Federation (EAFF), Vice-President of the Kenya National Federation of Agricultural Producers (KENFAP).

Raymond Labrousse
Geographer and pedologist, specializing in sustainable management of natural resources in tropical areas.

Cécile Lapenu
Agroeconomist, Executive Secretary of the CERISE (Comité d'échange, de réflexion et d'information sur les systèmes d'épargne-crédit) network, which brings together five French organizations working in support of microfinance institutions in southern countries (CIDR, CIRAD, GRET, IRAM, and IRC-SupAgro).

Philippe Lavigne Delville
Anthropologist, former scientific director of GRET (Research and Technological Exchange Group), editor of several collective works on land management, including *Quelles politiques foncières pour l'Afrique rurale ?* (Karthala 1998), and several policy papers on land use policy in Africa (European Union guidelines, white paper for Coopération Française).

Jean-Pierre Lemelle
Former agroeconomist at AFD, responsible for agricultural issues in West Africa and Central Africa from 1994 to 1999, agency director in the Horn of Africa and Algeria from 2000 to 2005, and head of mission in the Strategy Department's Evaluation and Capitalization Division until 2009.

Anna Lipchitz
Agroeconomist, researcher at AFD, member of the international research group La régulation des marchés internationaux de produits agricoles, and participant in the European economist network, Trade Economists Network.

Bruno Losch
Economist, seconded to the World Bank, coordinator of the RuralStruc research program on the "structural dimensions of liberalization for agriculture and rural development" (World Bank, Coopération Française, IFAD). Previously researcher at CIRAD, where he was responsible for the Family Farm program. Author of works on Côte d'Ivoire, the coffee and cocoa sectors, and the humid tropics, and other subjects.

Ibrahim Assane Mayaki
Former Prime Minister of Niger, former Executive Director of the Rural Hub in Dakar. Chief Executive Officer of the New Partnership for Africa's Development (NEPAD) since January 2009.

Vatché Papazian
Agroeconomist, responsible for land use issues at AFD, formerly responsible for the rural sector in AFD agencies in Senegal, Burkina Faso, and Mali.

Jean-Pascal Pichot
Systems agronomist, director of the editorial board for *Cahiers d'études et de recherches francophones—Agricultures*, former researcher and department head at CIRAD.

Vincent Ribier
Agroeconomist, researcher at CIRAD, UMR Moïsa, professor at Paris 1 and Paris 11, specialist in agricultural policy, process analysis, and international trade negotiations.

Claude Torre
Agroeconomist, responsible for trade and development at AFD; previously technical assistant for the Ministry of Cooperation (Niger, Guinea), then head of mission and cooperation attaché at the Ministry of Foreign Affairs (France, Morocco).

François Traoré
Burkinabé farmer, former President of the National Union of Cotton Producers of Burkina Faso, President of the Association of African Cotton Producers.

Betty Wampfler
Socioeconomist, professor, and researcher at the Institut des régions chaudes, SupAgro Montpellier, previously responsible for research on rural financial services at CIRAD.

Acknowledgments

This book was brought into existence under the guidance of a committee of the Agence Française de Développement (AFD), chaired by Jean-Michel Debrat and composed of Michel Jacquier, Jean-Yves Grosclaude, Jean-Bernard Véron, Bernard Esnouf, Vincent Joguet, Thomas Mélonio, Yves Jorlin, Caroline Contencin, and Henry de Cazotte.

The advice and comments of Bruno Losch, Georges Courade, Dominique Gentil, Valentin Beauval, Marie-Rose Mercoiret, Philippe Jouve, Jean Benhamou, Michel Griffon, Pierre Jacquet, Marc Bied-Charreton, Guillaume Ernst, François Peyredieu du Charlat, Georges d'Andlau, Jean-Pierre Prod'homme, René Billaz, Marc Dufumier, Jean-Marie Cour, Roland Pourtier, Frédéric Sandron, Jean-Pierre Chauveau, Samir Amin, Bruno Vindel, Freddy Destrait, Patrick Delmas, Roger Blein, and Bernard Bachelier provided valuable assistance at various stages of the book's development or during the preceding phase, involving a piece on "Agence Française de Développement and the future of the African family farm."

The preliminary version was submitted to Philippe Hugon and Jacques Brossier prior to publication.

A number of persons reviewed individual chapters, especially Alfred Schwartz, Jean Pichot, Roland Pourtier, Chantal Bernard, Monique Devèze, Dominique de la Croix, Valentin Beauval, Anne Lothoré, Christophe Jacqmin, Anne Legile, Patrick Delmas, Denis Pesche, Christian Agel, Guy Faure, Jean-Pierre Olivier de Sardan, François Rossin, Alain Maragnani, Jean-Bosco Bouyer, Christian de Gromard, Bruno Leclercq, Patrice Garin, Philippe Grandjean, Marie-Hélène Dabat, Patrick Gubry, Claude Remuzat, and Denis Loyer.

Finally, this book owes a great debt to the authors of all the different chapters who agreed to contribute to this collective effort; to Anne Perrin and Daouda Diagne for part 3; to African farmers and officials contacted in the field; to departed individuals who ably collaborated with them, Jacques Moineau, Michel Penent, and Jacques Alliot; and to all our colleagues at AFD, the Inter-Network Rural Development Association, CIRAD, and various French and international aid agencies that have made a commitment to rural development in Africa.

Abbreviations

ACP	African, Caribbean, and Pacific countries
AFD	Agence Française de Développement (French Development Agency)
AFDI	Agriculteurs Français et Développement international (French Farmers and International Development)
AGRA	Alliance for a Green Revolution in Africa
AMIRA	Amélioration des méthodes d'investigation en milieu rural africain (Improving the Methods of Investigation in Rural Africa)
AOPP	Association des organisations professionnelles paysannes (Association of Farmers' Trade Organizations) (Mali)
APAD	Association pour la promotion d'une agriculture durable
APROCA	African Cotton Producers Association
APU	Agricultural Policy of the Union
AUF	Agence universitaire de la francophonie
BMZ	Bundesministerium für wirttschaftliche Zusammenarbeit und Entwicklung
CAP	Common Agricultural Policy (European Union)
CECAM	Caisse d'épargne et de crédit agricole mutual
CEPED	Centre population et développement
CERISE	Comité d'échange, de réflexion et d'information sur les systèmes d'épargne-crédit
CET	common external tariff
CGDA	Conseil général du développement agricole
CICC	Interprofessional Coffee and Cocoa Board
CICRED	Committee for International Cooperation in National Research in Demography
CIDR	Center international de développement et de recherche
CIF	Centre d'innovation financière

CILSS	Comité permanent inter-Etats de lutte contre la sécheresse dans le Sahel (Permanent Inter-State Committee for Drought Control in the Sahel)
CIRAD	Centre de coopération internationale en recherche agronomique pour le développement (Center for International Cooperation on Agricultural Research for Development)
CNCR	Conseil national de concertation et de coopération des ruraux
CNIA	National Interprofessional Groundnut Board
CPRC	Chronic Poverty Research Centre
CRDI	Centre de recherche pour le développement international
CSS	Compagnie sucrière du Senegal
CVECA	Caisse villageoises d'épargne crédit autogérées
DGCID	Direction générale de la coopération internationale et du développement
EAFF	Eastern Africa Farmers Federation
ECOWAP	Agriculture Policy of the Economic Community of West African States
ECOWAS	Economic Community of West African States
EDI	Etudes et documentation internationales
ENDA (and ENDA Diapol)	Environmental development action in the developing world
EPA	economic partnership agreement
EU	European Union
FAC	Fonds d'aide et de coopération
FARM	Fondation pour l'agriculture et la ruralité dans le monde
FAO	Food and Agriculture Organization
FIDES	Fonds d'investissement pour le développement économique et social
FO	farmers' organization
FONGS	Federation of Senegalese Nongovernmental Organizations
GDP	gross domestic product
GDS	Grands domaines du Sénégal
GMDs	genetically modified organisms
GRET	Research and Technological Exchange Group
HIPC	Heavily Indebted Poor Country
ICCO	International Cocoa Organization
IER	Institut de l'économie rurale
IFAD	International Fund for Agriculture Development
IFORD	Institut de formation et de recherche démographique
IFPRI	International Food Policy and Research Institute
IIED	International Institute for Environment and Development
IMF	International Monetary Fund

INRA	Institut national de la recherché agronomique
IRAM	Institut de Recherche et d'Applications des Méthodes de développement
IRAT	Institut de recherché en agronomie tropicale
IRC	Institut des régions chaudes
IRD	Inter-réseaux Développement rural
IRD	Institut de recherche pour le développement
ISRA	Institut sénégalais de recherches agricoles
KENFAP	Kenya National Federation of Agricultural Producers
LARES	Laboratoire d'analyse régionale et d'expertise sociale
LIFDC	low-income food-deficit countries
LOASP	agro-sylvo-pastoral framework law
MAE	Ministère des affaires étrangères (Ministry of Foreign Affairs)
MDG	Millennium Development Goal
MINREST	Ministère de la recherche scientifique et technique
NEPAD	New Partnership for Africa's Development
NGO	nongovernmental organization
ODA	official development assistance
OECD	Organisation for Economic Co-operation and Development
OP	Organisation de producteurs (producer organization)
OPA	Organisation professionnelle agricole (professional agricultural organization)
ORSTOM	Office de la recherche scientifique et technique outre-mer
PASAL	Programme d'appui à la sécurité alimentaire
PO	producer organization
PROPAC	Plateforme régionale des organisations paysannes d'Afrique centrale
R&D	research and development
REDD	Reducing Emissions from Deforestation and Forest Degradation
ROPPA	Réseau des organisations paysannes et de producteurs de l'Afrique de l'Ouest (West African Network of Farmer and Producer Organizations)
SACAU	Southern African Confederation of Agricultural Unions
SAP	Structural Adjustment Program
SAR	Special Administrative Region
SNV	Netherlands Development Organization
SOCAS	Société des conserves agricoles du SENEGAL
UNCTAD	United Nations Conference on Trade and Development
UNPCB	Union nationale des producteurs de coton du Burkina Faso
WALTPS	West Africa Long-Term Perspective Study
WAEMU	West African Economic and Monetary Union
WTO	World Trade Organization

Introduction

Jean-Claude Devèze

In the face of mounting pessimism over the plight of Africa, further fueled by the proliferation of conflicts in Sub-Saharan Africa and the 2008 rioting caused by hunger, the need to ponder the future of African farmers and the role of agriculture in future development seems clear. Of the 750 million inhabitants of Sub-Saharan Africa, two-thirds (500 million) live in rural villages of fewer than 2,000 inhabitants where farming and stockraising continue to be the main occupations. In the great majority of cases, this work is performed by family units characterized by a variety of social structures. Alongside family farms, agribusiness—which still to some extent reflects a colonial past (Dufumier 2004)—plays only a marginal role in employment in most countries, except in South Africa where 5,000 large farms employ close to half the agricultural population and produce 95 percent of all output reaching the market (Nieuwoudt and Groenwald 2003).

This book uses the terms *family farms* and *smallholder agriculture* interchangeably. Family farming may be defined as modes of organizing life and production that revolve around the tight bonds that exist between social and economic activities, family structures, and local conditions (village lands, group membership); this may also be referred to as smallholder agriculture to underscore its deep local connections. Agribusiness, by contrast, is characterized by heavy reliance on agricultural employees and investment capital.

The Food and Agriculture Organization estimates the agricultural population in Sub-Saharan Africa (excluding South Africa) for 2003 to be 409 million. Two important points deserve particular attention: first, the size of this population; and second, its continued growth (14 percent between 1995 and 2003) as a result of high birth rates in the countryside, notwithstanding migrations to cities and foreign countries. This population growth raises important questions about the future of ever more numerous farm families with many children. Another important demographic issue to be analyzed is the evolution of the relative number of farmers in the population (box I.1).

BOX I.1

Differences between Countries Based on Relative Numbers of Agricultural Workers

In the 47 countries of Sub-Saharan Africa and the Indian Ocean for which the necessary demographic data are available, major distinctions can be drawn about how reliant they are on agriculture, based on two criteria described by Bruno Losch in chapter 2, namely, the percentage of decrease in the relative number of agricultural workers between 1961 and 2005, and the relative weight of agriculture among the working population in 2005.

- The relative number of agricultural workers fell by more than 50 percent in seven countries, topped by South Africa, and followed, in order, by Mauritius, Cape Verde, Gabon, Nigeria, Swaziland, and Botswana. In all these countries, farmers make up less than 50 percent of the working population.
- Seven countries were in the middle range, with a relative decrease of between 50 and 30 percent. In order, these countries were Côte d'Ivoire, Namibia, Republic of Congo, Mauritania, Benin, Cameroon, and Sudan.
- The 33 remaining countries, ranging from Togo to Burkina Faso, saw their relative number of agricultural workers fall by less than 30 percent. In all these countries, farm workers account for a majority of the working population.

The broad diversity of ecological systems, associated or dissociated activities (agriculture, forestry, stockraising, fisheries, fish farming), population densities (see map, p. vi), means of adaptation to natural constraints (for example, recourse to fallow periods in areas where land is still abundant), methods of social organization, and the like have resulted in a multitude of land systems. Specific to the local context, these systems reflect the choices and dynamics of a particular community in dealing with its natural environment, pressures on the land, and external factors. The diverse nature of African agriculture is a source of wealth because of all the potential it offers, but it is also a source of difficulty given the complexity of the situations encountered. In-depth knowledge of local realities is thus necessary to avoid falling back on ready-made solutions.

With low worker productivity (Mazoyer and Roudart 1997), most Sub-Saharan family farms cultivate small fields, mostly by hand—animal traction remains a minority practice and mechanization is rare—and women and children supply a large part of the manual labor. The use of fertilizers (13 kilograms a hectare of cultivated area), which is much lower than everywhere else in the world, is stagnant; only 4 percent of fields are irrigated, and improved seeds are used on just 24 percent of the total area planted in grain crops (World Bank 2007).

Depending on climate conditions and specific circumstances, these farmers manage with greater or lesser success to feed their own families and low-income rural and urban populations. The average daily food intake in the region is reported to be 2,400 calories, compared with 4,000 calories in OECD countries; moreover, this diet—93.6 percent of which is plant-based—lacks lipids and protein. In addition, while food imports have changed little in proportion to total imports in value (15 percent), grain imports and aid tripled between 1974 and 2000 and continue to go up (Hugon 2006). Last, the new context of rising agricultural prices poses the risk of greater problems in feeding Africa's poor (Diouf and Severino 2007). This raises a key question: is Sub-Saharan agriculture capable of becoming more proficient in the future at feeding an ever-growing population at a time when agricultural price tensions are increasing worldwide?

The growing population density places pressure on the land, on pasturage, and on wood resources in certain areas and thus creates tensions that may degenerate and become a source of conflict, as in Rwanda, Côte d'Ivoire, Kenya, or Darfur. In these conditions, how can Sub-Saharan Africa successfully reconcile population growth, sustainable management of its natural resources, and the peaceful coexistence of different groups? This is a second key question, and it bears on the future of millions of rural youth looking for work each year (Chauveau 2005) and on the question of alternatives that can be offered to young migrants who head for the city and find only informal and poorly paid work (Giordano and Losch 2007).

Given the all-too-frequent tendency to abandon the countryside to its fate, the task of improving living standards and conditions in rural areas and the necessity of strengthening the dynamics of regional development now emerge as critical concerns in numerous works about the future of Sub-Saharan Africa.

One final question that arises is cross-cutting in nature: What is the future of African family farms that must compete against much more productive farms that are better organized and receive major support of all kinds?

Despite more than 50 years of independence and development assistance, the future of African farm families and their children is increasingly worrisome. Yet agricultural and rural Africa is changing, as shown by various authors who have sought to understand the process now under way. How does one advance from the situation portrayed in *L'Afrique noire est mal partie* by René Dumont or *L'agriculture africaine en réserve* by Philippe Couty to African farms of the future? How does one best steer farms that are on a path to marginalization toward a truly dynamic rural economy? How can one best meet *the challenges for African agriculture*?

This book, in attempting to take into account the processes under way, adopts an approach that looks at transitions and how best to steer them, which of course means it is essential to clearly specify the transitions that are involved. In the 1970s the working group on Improving the Methods of Investigation in Rural Africa (AMIRA) studied the movement of rural societies toward capitalism and the risks that this transition entailed, particularly in relation to changes in the ways of making decisions. The World Bank, in its 2008 annual report, speaks of social transition in connection with building the skills for guiding agricultural and rural activities, and also of transition toward a market orientation so as to better respond to the demand for agricultural and nonagricultural products. In chapter 2, Bruno Losch examines the transitions of rural economies and how they contribute to developing a continent that consists mainly of agriculture-based countries. Some authors (Obaid and Severino 2007; Cour 2007), wagering on urban dynamics, present the growing nonagricultural population that will need to be fed as an opportunity for African agriculture; the main question then becomes: "How is Africa going to manage the transition from rural to urban?" Other authors (Courade and Devèze 2006; Jouve 2006) emphasize the importance of land reforms and agricultural revolutions if the planet is to be fed through sustainable development (Griffon 2006). The issue is thus to clarify which transitions of agrarian societies and rural economies will enable them to meet the challenges of the future.

Accordingly, the first part of the book seeks to present these sizable challenges, with special attention to three key challenges:

- The demographic challenge, given the continued growth of the agricultural and rural population and the substantial migrations
- The economic challenge, in terms of the place of agriculture in development, given the international context and the opening up of markets
- The environmental challenge, involving the use of natural resources and ecosystem management, in connection with land pressure and climate contingencies and change

The second part of the book presents five fields of action that can be incorporated into government policies to promote successful agricultural transitions. The following priorities were identified: improve and safeguard systems of land tenure, strengthen innovation processes and support mechanisms, capture regional markets for food products, develop financing for agriculture, and promote human capital.

The final part of the book gives the platform to four African officials who present their vision of the future of agriculture, along with their ideas about fields of action that need to be promoted and conditions that need to be satisfied in order to meet the challenges facing agriculture.

The conclusion summarizes the key messages to be drawn from this collaborative work.

A few points need to be clarified concerning the choices made in putting this book together. First, with respect to the geographic area involved, the focus was placed on the countries of Sub-Saharan Africa, including South Africa, and the islands of the Indian Ocean, with special attention given to Madagascar. Next, attention was given not just to agriculture but also to other important factors influencing the lives of farmers, such as the increased tendency to engage in multiple activities (already some 40 percent of the income of rural households reportedly comes from activities that are not strictly agricultural) and the ever greater overlap between cities and countryside. Finally, the fact that this book focuses specifically on agricultural development should not lead the reader to forget the importance of links between rural and urban development.

Bibliography

Chauveau, J.-P. 2005. "Les jeunes ruraux à la croisée des chemins." *Afrique contemporaine* 214: 15–33.

Cour, J.-M. 2007. "Peuplement, urbanisation et développement rural en Afrique subsaharienne: un cadre d'analyse démo-économique et spatial." *Afrique contemporaine* 223/224.

Courade, G., and J.-C. Devèze. 2006. "Des agricultures africaines face à de difficiles transitions." *Afrique contemporaine* 217.

Couty, P. 1996. *Les apparences intelligibles, une expérience africaine*. Arguments, Paris.

Devèze, J.-C. 1996. *Le Réveil des campagnes africaines*. Karthala, Paris.

Diouf, J., and J.-M. Severino. 2007. "Feeding Africa." *International Herald Tribune,* October 20.

Dufumier, M. 2004. *Agricultures et paysanneries des Tiers mondes*. Karthala, Paris.

Dumont, R. 1962. *L'Afrique noire est mal partie*, Le Seuil, Paris. [English edition: *False Start in Africa*, 1966, Praeger, New York].

Grain de sel 23. June 2003. Report on "Agricultures familiales, de qui et de quoi on parle?"

Giardano, T., and B. Losch. 2007. "Transition: Risques d'impasse." *Courrier de la Planète.*

Griffon, M. 2006. *Nourrir la planète*. Odile Jacob, Paris.

Hugon, P. 2006. *L'économie de l'Afrique*. La Découverte, Paris.

Jouve, P. 2006. "Transition agraire: la croissance démographique, une opportunité ou une contrainte?" *Afrique contemporaine* 217.

Mazoyer, M., and L. Roudart. 1997. *Histoire des agricultures du monde*. Le Seuil, Paris.

Nieuwoudt, L., and J. Groenwald. 2003. *The Challenge of Change: Agriculture, Land and the South African Economy.* University of Natal Press, Scotsville.

Obaid, T., and J.-M. Severino. 2007. "L'urbanisation, risque ou chance ?" *Le Monde*, June 28.

World Bank. 2007. *World Development Report 2008: Agriculture for Development.* Washington, DC.

Part 1

African Agriculture in the Face of Multiple Challenges

Among the many challenges that appear to hold great importance for the future of agriculture in Sub-Saharan Africa, three have been judged most critical, namely, the demographic challenge (chapter 1), the economic challenge (chapter 2), and the environmental challenge (chapter 3). Sub-Saharan Africa is, in fact, the only subcontinent that continues to have a consistently high population growth rate (2.3 percent), a fertility rate twice the world average, and more than 60 percent of its population living in rural areas. In addition, Sub-Saharan Africa faces a specific economic challenge: how to bring about a process of sustainable economic growth, made necessary by the rapid increase in its population, with a somewhat uncompetitive agricultural base, all within a highly competitive economic environment at the international level. Last, the sheer magnitude of tropical Africa's potential biomass production, which could make it an agricultural giant along the lines of Brazil, cannot obscure the region's vulnerability: resources that must be shared under increasingly contentious conditions; difficult control of the exploitation of ligneous resources and reconstitution of forest resources; soil erosion and the fragility of projected landscapes following deforestation; water control in the context of climate change—the list goes on.

But all this hardly negates the importance of the social, cultural, and political challenges discussed in chapter 4. All these challenges are intertwined, which leads the various analysts of the agricultural and food situation to highlight some challenges rather than others.[1] The focus is thus on clarifying the way in which these challenges are addressed, by proposing that they be subsumed under the three following goals: develop the natural potential, which is substantial, to feed first and foremost Africans; promote the availability of human capital on family farms; and make sure that these efforts to develop and promote are lasting, widespread, and relevant to the entire economy and society.

Note

1. In *Le Monde* of May 11–12, 2008, World Bank president Robert B. Zoellick discusses all facets of the new world deal on food policy: "This new deal should focus not only on hunger and malnutrition, access to food, and sources of supply, but also on their interconnections with energy, productive yields, climate change, investment, the marginalization of women, economic growth and resistance, and more." Michel Barnier, former French Minister of Agriculture and Fisheries, points out that, through food, climate, environmental, and commercial challenges, agriculture continues to shape our existence (www.parlonsagriculture.com).

Chapter 1

The Demographic Challenges

Francis Gendreau

Rural communities in Africa face formidable challenges: a consistently high population growth rate (2.3 percent), a fertility rate twice the world average, and urbanization that does not absorb all of the surplus population. Social relationships, technological changes, the status of women, and the education of rural youth will play a key role in the demographic transition under way, a transition that is especially unpredictable in the agricultural environment.

The demographic characteristics of Sub-Saharan Africa are quite different from those of the other regions of the world, particularly with respect to strong population growth that will no doubt continue for a number of decades. These characteristics must be examined not in isolation but in conjunction with all the cultural, social, political, economic, and environmental aspects of development. Only then will we understand the population dynamics and the social transformation mechanisms at work that need to be linked to demographic behavioral changes.

In addition, the region has in recent decades been experiencing economic and political turmoil that has slowed the pace of economic and social progress, resulting at times in a standstill and even, in some cases, a decline in living standards. Discourses, which adopt an oversimplified interpretation of Malthus's theory, too often tend to attribute these difficulties to demographics on the basis of a simple cause and effect relationship. We have not mirrored this approach but instead examine the demographic changes in their socioeconomic context, fully cognizant of the fact that complex interactions link population and development (Véron 1994).

It is against this backdrop that this chapter has been prepared.

The first part provides a demographic picture of Sub-Saharan Africa, reviewing the current situation and recent trends and then the prospects for the future: population numbers and the reasons for population movements that modify these numbers at the national level (international migration, fertility, mortality, and the ensuing growth). The economic, social, and political aspects of these dynamics, as well as their causes and consequences, are briefly examined. The specific characteristics of rural populations are highlighted whenever necessary.

The second part then specifically examines the issue of rural populations and their sociodemographic challenges. To this end, it briefly discusses the links between population and development and then focuses on the three key areas that result from the characteristics identified in the first part, namely, urbanization, rural population growth, and rural youth. An annex discusses necessary population policies.

The Overall Demographic Picture

It is useful to begin with a brief review of some specific traits of demographics in Sub-Saharan Africa.[1]

Even if we have to resort to oversimplification by disregarding all the characteristics that are peculiar to each country (Sub-Saharan Africa is not a homogeneous block!) and the geographic differences that are sometimes significant within a particular country (between regions, and between urban and rural areas), we will endeavor to summarize this demographic situation in five broad statements before examining the prospects for future trends.[2]

750 Million Inhabitants in 48 Countries

Sub-Saharan Africa is now home to some 750 million people—almost 12 percent of the global population—distributed among 48 countries.[3] (A list of these countries appears in table 1.1.[4]) These 48 countries vary considerably in size. On the one hand are Nigeria (which, with a population of over 130 million people, is the ninth most populous country in the world) and a number of big countries such as Ethiopia (77 million), the Democratic Republic of Congo (58 million), and South Africa (47 million), to name but a few with over 40 million inhabitants, while on the other hand, there are "microstates" that include countries with fewer than 600,000 inhabitants, such as the Seychelles (81,000), São Tomé and Principe (157,000), Equatorial Guinea (504,000), and Cape Verde (507,000).

The average population density is not very high (25 persons per square kilometer). The colonial-era mindset with respect to an "underpopulated Africa" (in 1960, the average population density was only 9 people per square kilometer) bears recalling, because at that time underpopulation was perceived to be an obstacle to development. However, the average varies considerably in each country and among countries. At one extreme Botswana, Central African Republic, Chad, Gabon, Mauritania, and Namibia have an average population density of fewer than 10 people per square kilometer, while at the other

Table 1.1 Population and Urbanization Rates (Mid-2005)

Country	Population (millions)	Urbanization rate (%)	Cultivated surface area (hectares per farmer)
Angola	15.9	37	0.39
Benin	8.4	46	0.56
Botswana	1.8	53	0.55
Burkina Faso	13.2	19	0.32
Burundi	7.5	11	0.21
Cameroon	16.3	53	0.88
Cape Verde	0.5	58	—
Central African Republic	4.0	44	0.75
Chad	9.7	26	0.55
Comoros	0.6	36[a]	—
Congo, Dem. Rep.	57.5	33	0.26
Congo, Rep.	4.0	54	0.12
Côte d'Ivoire	18.2	46	0.48
Djibouti	0.8	85	—
Equatorial Guinea	0.5	50	0.72
Eritrea	4.4	21	0.18
Ethiopia	77.4	16	0.20
Gabon	1.4	85	0.97
Gambia, The	1.5	26	0.17
Ghana	22.1	46	0.39
Guinea	9.4	37	0.11
Guinea-Bissau	1.6	36	0.30
Kenya	34.3	42	0.20
Lesotho	1.8	18	0.46
Liberia	3.3	48	0.19
Madagascar	18.6	27	0.26
Malawi	12.9	17	0.19
Mali	13.5	34	0.26
Mauritania	3.1	54	0.15
Mauritius	1.2	44	—
Mozambique	19.9	38	0.23
Namibia	2.0	34	0.71
Niger	14.0	23	0.38
Nigeria	131.5	48	0.86
Rwanda	9.0	22	0.17
São Tomé and Principe	0.2	38	—

continued

Table 1.1 *continued*

Country	Population (millions)	Urbanization rate (%)	Cultivated surface area (hectares per farmer)
Senegal	11.7	51	0.34
Seychelles	0.1	50	—
Sierra Leone	5.5	40	0.20
Somalia	8.2	36	0.16
South Africa	47.4	58	2.11
Sudan	36.2	41	0.68
Swaziland	1.0	24	0.54
Tanzania	38.3	38	0.13
Togo	6.1	36	0.89
Uganda	28.8	12	0.37
Zambia	11.7	37	0.73
Zimbabwe	13.0	36	0.36
Total	**750.4**	**37**	**0.40**

Source: Population figures are from United Nations 2005; urbanization rates are from United Nations 2004; and data on cultivated land come from Bot, Nachtergaele, and Young 2000 (farming population was updated in 2000).
a. Includes Mayotte.
— = Not available.

extreme are small, very densely populated countries such as Burundi, Comoros, Mauritius, and Rwanda, with more than 200 people per square kilometer.

When the issue of population density is raised, the focus tends to be on the rural population, which is covered in the second part of this chapter. We simply wish to note at this juncture that certain inland regions of a country may have very densely populated rural areas, such as Bamiléké country in western Cameroon, the Mandara Mountains in northern Cameroon, Ibo country in southeastern Nigeria, or Kabiyé country in northern Togo, which could result in strong demographic pressure given the quality of the soil, climate conditions, and cultural practices. However, a number of countries have desert regions (such as the Sahara and the Kalahari), where populations are low. For example, the Gao, Kidal, and Timbuktu regions of Mali, which cover 66 percent of the country, are home to less than 10 percent of the population and have a population density of 1 inhabitant per square kilometer.

Major Migration Flows Are Primarily Intracontinental

The migratory currents from Sub-Saharan Africa to the rest of the world are relatively unimportant, although they generate a plethora of media coverage. There is a small migration flow to North America; an example is an early emigration flow from Cape Verde to the northeastern coast of the United States.[5] Immigrants primarily head to Europe. In France, for example, the immigration of "the people from the river valley" (Soninké and Tukulor people living along the Senegal River that runs through Mali, Mauritania, and Senegal) in the postwar era is well documented. In connection with immigration restrictions, this immigration has declined since 1974 and changed in character; the noria (shifting migration) of young, short-stay immigrants has been replaced by families intending to settle on a more permanent basis. More generally, border closures have triggered illegal migration to European countries, especially France, Spain, and Italy. However, these migrations flows were negligible from a quantitative standpoint (although, for the countries of origin, they can have a major economic impact).[6]

International migration in Sub-Saharan Africa is thus, to a large extent, intracontinental. Limited quantitative data on this phenomenon are available although certain long-standing trends are well known: from the Sahel (Burkina Faso, Mali, Niger, for example) to coastal countries (such as Côte d'Ivoire, Ghana, and Nigeria), and from countries bordering South Africa (such as Botswana and Mozambique) to South Africa, for example.[7]

Thus, in 1998 (before the onset of the Ivorian crisis), the population of Côte d'Ivoire included 2.2 million immigrants (foreign-born persons), that is, 14 percent of the total population. This number of immigrants was almost as high as that of foreigners (4 million, or 26 percent of the population), who are not all immigrants—far from it, in fact, since close to half of them were born in Côte d'Ivoire. This country also possesses another unique trait: contrary to general observations, a significant portion of these foreigners (over 40 percent) were living in rural areas.

However, in addition to these "voluntary" migrations, the refugee movement, another tragic dimension of migration, also bears noting. Sub-Saharan Africa is home to 2.6 million refugees, that is, 20 percent of the total number of refugees identified around the world.[8] These refugees are to be found chiefly in Chad, the Democratic Republic of Congo, Kenya, Tanzania, and Uganda. They originate mainly from Angola, Burundi, the Democratic Republic of Congo, Liberia, Somalia, and Sudan. Moreover, large numbers of displaced persons can be found in certain countries, owing to wars or forced displacement (Liberia, Somalia, and Sudan, for example). Even if the international community comes to the aid of these populations, they are often a considerable burden for poor states and contribute to their disorganization.

Mortality Remains High and in Chaotic Decline

Although the mortality rate has generally been trending downward for many years, it is still quite high, especially among young children. Moreover, the decline has been chaotic, with periods where it has slowed, stalled, or even reversed, in tandem with the myriad problems faced by most Sub-Saharan African countries: the stagnation or decline in living standards, the deterioration of health systems, and the emergence of new diseases, especially AIDS, for example. Considerable progress in combating mortality still needs to be made: life expectancy at birth is just 46 years (the global average is 65 years), the under-five mortality rate is 173 per 1,000 live births (it is 86 per 1,000, or half that rate, in the rest of the world), and the infant mortality rate is 101 per 1,000 (the global average is 57 per 1,000).

These rates are particularly high in countries that are experiencing or have experienced unrest or wars (Liberia, Sierra Leone, and Somalia), and in Sahelian countries (Burkina Faso, Chad, Mali, and Niger), where more than one in five children die before the age of five.

All studies show that mortality is systematically higher in rural than in urban areas, and the difference can be greater than 50 percent in countries such as Madagascar, Malawi, and Niger. This difference can be attributed to various reasons, including poverty and harsher living conditions, a lower level of education, and more difficult access to health care. A comparable situation exists in the most disadvantaged neighborhoods of big cities where unsanitary conditions, overcrowded homes, the absence of running water, and other such challenges constitute a fertile ground for high mortality. Thus, for example, the under-five mortality rate (180 per 1,000) in shantytowns in Ethiopia is almost double that of other areas in the big cities (95 per 1,000).

The HIV-AIDS epidemic is having tragic consequences with respect to mortality. The United Nations estimates that in the 40 Sub-Saharan African countries hardest hit by the epidemic, life expectancy at birth fell from 48 years in 1990–95 to 46 years in 2000–05, whereas, without the epidemic, it was projected to increase from 51 years to 54 years during the same periods.[9] The situation is even more dire when consideration is given to the four hardest-hit countries, namely, Botswana, Lesotho, Swaziland, and Zimbabwe, where life expectancy at birth plummeted from 57 years to 37 years during the periods in question.

Fertility Is Still Far from Being Brought under Control

In traditional rural societies that use nonmechanized farming methods, having numerous offspring is an asset for economic production within the family unit (more hands available to work on the land),[10] a source of social prestige (offering the possibility of strengthening alliances through marriage), and a safety net in old age (children are the best "old-age insurance" when social security coverage is unavailable). In addition, socioeconomic structures encourage the birth of children, especially because a high infant mortality rate prompts women and couples to plan for more children to replace deceased ones.

Modernization of the economy (monetarization) and urbanization will result in a lower fertility rate, owing to improvements in living standards and level of education, the decline in the infant mortality rate, and an increase in the age at marriage.

However, the fertility rate remains high in Sub-Saharan Africa: the average number of children per woman in this subregion is 5.5, whereas the figure is 2.7 for the rest of the world. Because fertility transition in Sub-Saharan Africa is just beginning, the situation varies considerably from country to country:

- A number of countries still have a high fertility rate, and available data do not indicate the beginning of a decline (Chad, Mali, Niger).
- In several countries, the number of which is steadily increasing, the fertility rate has now begun to fall (Benin, Cameroon, Eritrea, Madagascar, Malawi, Senegal).
- These countries are gradually joining the group of countries where a decline in fertility began, in some cases, several years ago (all the countries in Southern Africa, Ghana, Kenya, Zimbabwe).
- Only Mauritius and the Seychelles, two Indian Ocean islands, are either in the final stage of their fertility transition or on the verge of completing it.

The reasons for these different trends can be attributed, among other things, to poverty levels, living conditions, the education level of women, the infant mortality rate, and access to contraception. With respect to this latter point, the prevalence of contraceptive use varies significantly from one country to the next, ranging from less than 5 percent in the Central African Republic, Chad, and Niger to over 40 percent in Namibia, South Africa, and Zimbabwe).[11]

As is the case with infant mortality, the fertility rate is still higher in rural than in urban areas. For example, in Ethiopia, a country where the gap is one of the widest, the rural fertility rate (6 children per woman) is 2.5 times higher than the rate in urban areas (2.4 children per woman).

The aforementioned factors explain this difference and are particularly relevant with regard to the prevalence of contraceptive use, which may be negligible in rural areas (it is lower than 1 percent in rural Chad and Mauritania), but exceeds 60 percent in urban areas in South Africa and Zimbabwe.

The same factors undoubtedly also explain why the fertility rate in working-class neighborhoods falls between the rate in rural areas and that of the other big city neighborhoods, as is evident for example in Burkina Faso, Cameroon, Chad, Mozambique, and Nigeria (UNFPA 2007).

The fertility rate appears to be just slightly lower among women in polygamous relationships. But the incidence of polygamy is generally low, with the exception of West Africa (Benin, Guinea, Togo), where polygamy rates among married men exceed 30 percent.

A Still Brisk Population Growth Rate, Despite the Onset of a Slowdown

The demographic transition recognizes that populations shift from a situation of a low growth rate resulting from a high fertility rate and a high mortality rate, to one where low population growth stems from low fertility and low mortality rates. During the transition period, the population growth rate peaks, because the mortality rate declines before the fertility rate does.

All African countries have begun their transition; in most cases, the decline in their mortality rate has been under way for some time. However, as we have seen, the situations vary considerably. While the mortality rate in a number of countries has fallen significantly, it continues to exact a heavy toll in others, and although some countries have already achieved a low fertility rate, others have seen almost no change in theirs. The result is a very wide range of growth rates, from below 1 percent (Southern African countries, Mauritius, Seychelles, and Zimbabwe) to over 3 percent (Benin, Burkina Faso, Chad, Niger, Somalia, and Uganda, for example).

However, as a whole, Sub-Saharan Africa has the highest population growth rate in the world—an average of 2.3 percent. This is attributable to the consistently high fertility rate, whereas, owing to the already long-standing decline in the fertility rate in most of the other regions in the world, the global population is currently growing only at a rate of 1.2 percent (and all developing countries at roughly 1.5 percent).

Furthermore, while the growth rate in all other regions in the developing world has generally been declining since at least the late 1960s, the rate in Sub-Saharan Africa only began to decline in the 1980s, from a peak of 2.9 percent during the 1980–85 period (the rate was 2.2 percent during the 1950–55 period).

A Scenario for Future Trends

Population prospects must always be viewed with caution. The ones given here, which were developed by the United Nations Population Division (United Nations 2005), are premised first and foremost on the implicit assumption that there will be no disasters (such as wars, famines, droughts, or floods) that could affect population trends.[12] They are based on the model of demographic transition whose use is expected to become more widespread.

However, in view of the diverse socioeconomic contexts, the extremely vast array of current demographic situations cited above, the uncertainties surrounding the speed with which the mortality and fertility rates will decline in the future, and the future scope of migration flows, it would seem advisable to view these prospects as one possible scenario rather than a prediction. We examine the assumptions made and the results they produce.

The Assumptions: Widespread Decline? The assumptions on which population prospects are based relate to mortality, fertility, and international migration.

We have already seen that there has been a historical downtrend in the mortality rate, despite occasional rate fluctuations and periods of reversal.[13] The United Nations is projecting an increase in life expectancy at birth in Sub-Saharan Africa, which is expected to climb from 46 years in 2000–05 to 64 years in 2045–50, an average gain of 2 years for each quinquennium. The range in life expectancy among countries is projected to be closer at the end of the period than at the beginning, that is, between 55 and 75 years, but with rates lower than 55 years in countries hardest hit by AIDS (Botswana, Lesotho, Swaziland, and Zimbabwe) and rates that could exceed 75 years in some places (Cape Verde, Comoros, and Mauritius).

The assumption made with respect to fertility is one of widespread control. This control is contingent upon several factors, such as an increase in age at marriage and, especially, the widespread use of contraception. It also requires a favorable socioeconomic context, where the infant mortality rate in particular has dropped, and where women have a higher level of education and enjoy an enhanced status and role. While the United Nations has projected that the global fertility rate will have virtually reached its replacement level in 2045–50 (2.05 children per woman), it is expected to remain at 2.6 in Sub-Saharan Africa. The range of rates is also projected to be tighter in this case, averaging between 1.9 and 3.3 children per woman, except in Mauritius and South Africa, where it could be lower, and in Burundi, Chad, Liberia, Niger, and Uganda, where it could be higher.

The future trend in international migration is predicated on an assumed general slowdown in migration. For all Sub-Saharan African countries, between 2005 and 2050, this assumption projects a total net emigration of 7 million people (while the population is expected to increase, with 940 million more births than deaths). International migration (which directly depends on the respective economic situation in the countries of origin and destination) is expected to play a very minor role in these prospects. It appears that no country has a high level of immigration and only a handful, namely Burkina Faso, Cape Verde, the Democratic Republic of Congo, Lesotho, Mali, Nigeria, and Senegal, have been identified as having a relatively high emigration rate (but that generally declines over time).

These three assumptions, like all assumptions, can, of course, be challenged. Is it truly realistic to envisage such a decline in the mortality rate given the current African context? Can one reasonably expect such a decline in the fertility rate while the decline is still faltering in a number of countries? Last but not least, is it not ambitious to imagine a slowdown in international migration, contrary to the very nature of migration flows, which societies have used since time immemorial as a means of adaptation? Can one conceivably accept the view that the population in Uganda will swell from today's figure of 29 million to 127 million in 2050, with an annual net emigration of only 2,000 persons over the next half century? Therefore, although the scenario outlined in this paper can be examined in broad terms, drawing conclusions from this scenario at the country level should be avoided.

Two Major Results: Reshuffling the Maps and "Relaxation." Based on these prospects, Sub-Saharan Africa is projected to have a population of 1.7 billion people by 2050, thus surpassing China (1.4 billion) and India (1.6 billion). This figure would be more than double that for 2005, whereas the global population is projected to increase by only 1.4 times. Sub-Saharan Africa would thus account for almost 20 percent of the global population, that is, close to the percentage that it represented in the early 17th century. Since that period, Sub-Saharan Africa's share of the global population has declined, owing to slavery and the triangular trade, and wars (including colonial wars); it hit its lowest level around 1920—on the order of 6 percent—before beginning to increase to reach today's rate of 12 percent and possibly 20 percent again in the future. This would bring about substantial changes to the global population map, and Sub-Saharan Africa's population share would increase significantly. What would the political and economic impact be?

The growth rate (which has begun to slow and will probably continue to do so) is expected to remain on the order of 1.3 percent in 2045–50 (compared to only 0.4 percent at the global level). "Demographic relaxation" would be indisputable were this rate of 1.3 percent to be compared with the current 2.3 percent, and even if this growth rate is still rapid because of the relatively recent nature of the decline in the fertility rate and the potential growth represented by the current youth population. Within Sub-Saharan Africa, the wide variety of situations previously cited is expected to produce vastly different growth rates. While a number of countries are projected to have a low population growth rate by 2050 (seven countries, Botswana, Lesotho, Mauritius, Seychelles, South Africa, Swaziland, and Zimbabwe, could have growth rates below 0.5 percent), a few others (Burundi, Chad, Republic of Congo, Guinea-Bissau, Liberia, Niger, and Uganda) could still experience growth rates above 2 percent. These differences in growth rates will alter each country's respective weight, which would necessitate a reshuffling of population maps, not only at the global level but also within Africa, which would thus have implications for age structures (see below).

The Rural Population and Its Challenges

Following the presentation of the major characteristics and trends, this chapter examines the accompanying changes in population structures and relates these changes to the socioeconomic transformations. We examine these issues with respect to rural population dynamics. Consequently, we first provide a number of factors pertaining to the thorny issue of the links between population and development. We then focus more specifically on urbanization, population growth, its territorial distribution, and its youth, all areas on which this chapter rightly places special emphasis.

Turmoil in the Links between Population and Development

As mentioned in the introduction, Malthus's theory (1798) has some influence on the views held by many political leaders who believe that rapid population growth is harmful to development: "The reality is that Africa's population is too big while its economic growth rate is too low."[14]

Yet, this theory has not yet been endorsed by scientists, who are more inclined to state that the links between population and development are part of complex and sometimes contradictory interdependent networks, and that population growth must be repositioned in its political, economic, and social context: the failure of agricultural development, the dominance of international trade by the more developed countries and their multinationals, the deterioration in terms of trade, the unjustifiable debt burden, the consumption explosion in rich countries, the rampant corruption, and so forth.[15] The complex nature of the phenomena and their interrelations does not lend itself to an overly simplistic generalization, which would hardly be a reflection of the prevailing situation.

The demographic transition process began in Europe in the late 18th and early 19th centuries, in a context of the "modernization" of Europe's economies and societies. Contrary to the Malthusian theory, these decades saw the simultaneous occurrence of robust economic growth and (relatively) strong population growth. Numerous other counterexamples can be cited, including the case of Côte d'Ivoire, where the "Ivorian miracle" (economic growth averaged close to 8 percent annually between 1960 and 1980) was achieved in a context of robust population growth (an annual average of around 4 percent over the same period). And the economic crisis began in Côte d'Ivoire at the very time that population growth had begun to slow down! There is thus no automatic cause and effect relationship between rapid population growth and weak development or between a population slowdown and development.

The Malthusian problematic has, nevertheless, been taken up by several international organizations and bilateral aid organizations with respect to Africa and its agricultural development, along with the population-poverty-environment "nexus": population growth would result in poverty and environmental degradation, with the latter exacerbating the former. Neo-Malthusians believe that the only way to escape from this vicious circle is to control the fertility rate.

However, trends in the three areas are a tangible reflection of the "responses" from societies to the changes they are experiencing. The analysis of the links between food security and population growth should thus critically examine the responses provided by societies to address the issue of increased food needs (Gendreau et al. 1991). These needs could indeed increase either on an ad hoc basis following a reduction in (or stagnation of) production, owing to various factors such as droughts or unrest, or in the medium term following population growth (demographic pressure). Various types of "demographic responses" therefore exist: increased mortality (famines), a decline in the fertility rate, and increased emigration. There may also be "economic responses" through increased production resulting from efforts to improve labor productivity and land yield,[16] or to technological breakthroughs. This theory, posited by Ester Boserup, refutes Malthus's theory. Boserup states that population growth is the key determining factor in the technological changes seen in agriculture (Boserup 1970).

The wide range of these responses, in terms of timing and location, is a clear indication of the need to take into account the geographic, historical, social, and cultural aspects of local contexts (Mathieu 1998). Consequently, the issue of the future of rural areas in West Africa where river blindness has been eradicated "cannot be raised simply in terms of population growth, farming methods, or environmental load capacity" (Quesnel et al. 1999). The authors show that repopulation of the space can be carried out in conditions where the use of resources can either weaken the environment or bring about sustainable development.

An Urbanization Rate That Is Still Low but Rapidly Rising

Historically, mobility has been a characteristic shared by almost all Sub-Saharan African countries. It is reflected in the various migration movements stemming from numerous strategies (individual and family) and involving, more often than not, significant reversals. In this regard, we previously referred to international migration flows, which are not a separate category but part of a broader picture that also includes internal migration flows: migration within rural areas (migration of people or colonization of new land), and especially urbanization (rural-urban migration). Doubtless in view of the immigration restrictions imposed by the industrial countries, migration of all types is generally looked on with mistrust. However, migration is one of the ways in which societies adapt to changes in their environment.

This observation is pertinent to urbanization, and a discussion of this phenomenon in a paper focusing on agriculture is necessary for three reasons:

- Movements (of population, money, food) between urban and rural areas are part of how an economy and societies operate; for city dwellers, relationships between cities and villages often remain strong.

- The modalities of urbanization clearly have an impact on rural areas, especially because there are various forms of migration, and movements can be reversed, as evidenced in return migrations from cities to villages, a phenomenon that has been studied, for example, in Côte d'Ivoire (Beauchemin 2001) and Cameroon (Gubry et al. 1996).
- While a percentage of the rural population does not engage in agricultural activity, many farmers reside in urban areas (urban and periurban farmers), where farming is more intensive than in rural areas.

Finally, the border between the urban and rural world is disappearing, and urban and rural development should be planned in tandem.

Urbanization in Sub-Saharan Africa is a recent phenomenon; in 1950 the urbanization rate was just over 10 percent. Today the rate has climbed to 37 percent (while the global rate is close to 50 percent). Again, there are considerable disparities among the various countries, with the rate ranging from below 15 percent in Burundi and Uganda to over 80 percent in Gabon (see table 1.1).

Urbanization has indeed been rapid over the past 50 years. Today, the urban population is growing at an annual rate of 3.9 percent. This rapid pace of urbanization is slowing down—the average annual rate was 4.8 percent between 1950 and 2005. During that period the urban population increased by more than 13 times, pushing the number of city dwellers from 20 million up to 278 million. In comparison, in Europe between 1860 and 1900, a period of intense urbanization related to industrialization, the annual urban population growth rate was only 2.3 percent.

Another characteristic is that urbanization is relatively diffuse in Sub-Saharan Africa. Almost 60 percent of the urban population lives in cities with fewer than 500,000 inhabitants. And in this subregion, very large cities are developing only very gradually: today, Sub-Saharan Africa has only one major urban area with over 10 million residents—Lagos (11.1 million)—its sole major urban area is among the 30 largest cities in the world.

Urbanization is a "heavy" trend in our societies and is expected to continue in Sub-Saharan Africa, despite the projected slowdown in pace. The United Nations estimates that by 2030 the urbanization rate in this subregion could exceed 50 percent.[17] The urban population is projected to grow by 2.9 percent each year during the 2025–30 period and by 3.4 percent between 2005 and 2030. In absolute terms, this would mean that the urban population is expected to increase by 2.3 times, soaring from 278 million in 2005 to 644 million in 2030.

This urbanization could be an asset (development and urbanization are two closely linked phenomena), provided cities serve as productive poles of development and markets in which small farmers can sell their products. A "virtuous" urbanization plan would, in effect, seek to ensure that rural development provides surplus production capable of supplying food to cities (thus increasing income for the small farmers) and that productivity gains result in the creation of a labor force that arrives in cities where business is expanding and where jobs in the secondary and tertiary sectors are available. This is not always the case in Sub-Saharan Africa, where migrants leave their rural environment, which lacks economic dynamism and essential infrastructure (such as schools, health centers, and roads), only to settle in poorly performing cities, the majority of whose residents are poor, and where very few job opportunities in the modern sector exist. African cities are traditionally not industrial centers, the exception being mining cities and cities in South Africa.[18]

Governments and municipalities are generally unable to meet the manifold needs of an increasingly expanding, albeit poor, population, especially with respect to services and infrastructure expenditures, such as housing construction, education and health services, urban transportation, supply of markets, sewage and waste disposal and treatment, and the supply of water and electricity. However, solidarity networks that facilitate several types of movement and urban integration have helped migrants settle and implement survival strategies in the informal sector. They are in a vulnerable position (at least during a certain period) and earn low wages, although these are higher than those they could earn in rural areas. In this respect, urban and periurban agriculture can play a key role in food supply, job creation, and income generation.

Continued Rural Population Growth and Population Redistribution

Owing to the rapid growth in total population and the still moderate level of urbanization, migration toward cities is not absorbing all of the surplus population resulting from the excess of births over deaths. Moreover, despite the rural exodus, rural populations in Sub-Saharan Africa continue to increase each year (at a current rate of 1.4 percent), unlike most of the other regions in the world where urbanization “is draining rural areas” (Europe: –0.6 percent; China: –0.8 percent).

This situation is expected to persist during the entire projection period and, in 2025–30, the rural population is expected to continue growing (0.5 percent). Between 2005 and 2030, the rural population is projected to jump from 473 million to 604 million, that is, an increase of “only” 1.3 times (whereas the urban population is expected to increase by 2.3 times).

By and large, rural population growth has led to increased pressure on land. In this regard, it would be more pragmatic to think in terms of cultivated land per farmer (and potentially cultivable land area) rather than in terms of density, according to calculations by the Food and Agriculture Organization (Bot, Nachtergaele, and Young 2000). An overview of current contrasts is available in the table 1.1. Cultivated land varies between 0.1 hectares and 1 hectare per farmer (except in South Africa, where the figure is 2.1 hectares), with an average of 0.4 hectares for all of Sub-Saharan Africa.

As we have seen, the population may use several possible strategies to address the issue of increased land pressure. These include low fertility, emigration, seasonal displacements, land clearance, increased cultivation by increasing work quantity or adopting new technologies (such as inputs, increased selection of seeds), crop diversification, or multiple occupations to minimize risk (including nonagricultural activities). Development will be contingent upon several factors, including:

- Social structures that can curb or support and promote behavioral change, particularly the expanded role of women in food production stemming from a decline in the relative share of working men owing to emigration to cities, and the necessary establishment of new social production relations between men and women.
- The "quality" of the populations and the workforce in terms of health status and level of education (general and technical training).
- Public policies in the various relevant areas, such as fertility control, education and health infrastructure, land tenure security,[19] technical training of small farmers, marketing networks and organization of markets, access to credit, and agricultural price setting.
- Economic growth and the distribution of investments among agricultural and nonagricultural sectors.

One of the most immediate responses is increased movements, "a factor to be taken into account with respect to agricultural dynamics and the development of rural spaces" (Quesnel 2001). This is reflected in population redistributions related to the search for land and access to employment (along new communication routes, on the outskirts of big cities, in spaces in which hydro-agricultural works are being executed, following forest exploitation that paves the way for agricultural colonization, and so forth). This movement must be taken into account in the discussion on sustainable farming systems (Marchal and Quesnel 1997). In rural areas with a relatively small population or more land, or both, the natural movement and rural exodus of populations will thus be supplemented by immigration from more densely populated or even over-populated areas, in view of the production systems implemented. For a number of reasons (historical,[20] medical, land-related, or other), the less densely populated areas may be seen, for example, as the West African valleys where river blindness has been eradicated (as previously mentioned), in the southwestern and central parts of Côte d'Ivoire, central Nigeria, the lowlands surrounding the Ethiopian highlands, and Kivu.

Given a growing population in need of food,[21] agriculture will long remain a key sector in African economies. A boost in agricultural production through an increase in land area (where possible), work productivity gains, and higher yields is imperative. Adaptation by small farmers is also necessary to enable them to meet growing demand from urban areas.[22]

However, this growth will create a surplus labor force. To ensure that this surplus labor does not significantly increase the number of potential rural migrants (leading to a further increase in urbanization or international emigration), it is necessary to develop nonagricultural activities (crafts, agrifood industry) and improve living conditions (housing, health, safe water, and the like) in rural areas. New land must also be developed and agricultural migration (see, for example, the operation pertaining to midwestern Madagascar) must be promoted. These prospects can be designed only in the context of proactive policies for land use planning and environmental preservation, and taking into account population dynamics.

Such policies are especially critical for very densely populated areas that could experience further population growth, particularly through inadequate emigration (for cultural or other reasons), and where access to land may pose difficulties for young people or immigrants, owing to the allocation method in place. Under these circumstances, social tensions could develop and escalate into unrest or even war. Several examples could be cited (Rwanda, Côte d'Ivoire) where demographic pressure has undoubtedly sometimes been a key component of conflicts.

All of these observations clearly show the need to simultaneously address development issues from both an urban and a rural perspective and the difficulties in achieving balanced development between the two worlds, as well as the need to rethink these issues: "When we talk about demographic transition, land programs, village land management, movement of people, in a territorial network that has been expanded to include multiple occupations, these are some of the approaches that are reaffirming the traditional concept of the rural dimension, which are positioned between consideration of the past and the future, development, and culture" (Marchal 1997).

These transformations are accompanied by a "social upheaval" (Marchal 1997) that emphasizes social distinctions, which could lead to some measure of "deagrarianization" in rural areas (Bryceson 2004), and which at times significantly modifies social relations.

A Still Young Population That Is Facing a Slow but Inevitable Aging Process

In addition to the slowdown in population growth, the demographic transition has a second major impact: an aging population.

Given the relatively recent nature of the decline in the fertility rate, the Sub-Saharan population is still young—44 percent of its population is under 15 years, compared with 28 percent in the rest of the world, while the 65-and-over age group still accounts for just 3 percent of the population, compared with over 7 percent in the rest of the world.

No value judgments should be made about this situation (for example, there are too many young people in Africa just as there are too many elderly persons in Europe). We will simply underscore one point: persons between the ages of 15 and 64, the working-age population ("potential workers"), account for just under half of the population and thus have responsibility for just over half of the population. This scenario differs significantly from the general global one, where just under two-thirds of the population are of potential working age and potentially responsible for just over one-third of the population.

In approximately 10 countries, young people (under 15 years) account for less than 40 percent, including Mauritius, the sole country with a youth population below 30 percent of its total population. The proportion of elderly persons, that is, persons ages 65 and over, varies between 2 percent and less than 4 percent, except in roughly 10 countries, where this figure remains below 5 percent; this figure, however, does not include Mauritius and Lesotho.

This population is projected to remain relatively young in the coming decades: the percentage of young people in this subregion is forecast to be 30 percent by 2050 (20 percent for the rest of the world), while its elderly population is expected to remain below 5 percent (16 percent for the rest of the world).

Given that the decline in the proportion of young people is much greater than the increase in the proportion of the elderly, the share of the working-age population is projected to increase during this 50-year period; this period is sometimes referred to as the "demographic bonus" or the "demographic dividend" (Bloom, Canning, and Sevilla 2002), which could be favorable to the economy because the responsibility of each (potential) worker for children and the elderly is decreasing. An analysis of the economies of several Asian countries whose working-age populations have grown reveals that this factor was instrumental in their rapid development.

This young age structure thus raises several societal questions:

- How will young people and the elderly be integrated into society and the work environment and what will their roles be?
- How will intergenerational relations develop within the family unit, where increasingly three or four generations now coexist?
- Will states take advantage of the favorable demographic bonus period to invest heavily in their youth? Will this be applied to youth desirous of remaining in rural areas and becoming involved in agricultural activities as well as those who opt to migrate to the city?

The crisis now experienced by several Sub-Saharan African countries and the constraints imposed on social budgets by structural adjustment programs are reflected in the major challenges related to education, access to health services, and employment for young people. As a result, the model for social promotion used in previous decades has been called into question; government too often perceives the high proportion of young people more as a source of instability than as an asset for future economic growth. However, this potential demographic bonus will only be realized if this youth population is prepared to become an efficient adult population from an economic standpoint and a responsible citizenry from a political one, which requires an effective education system accessible to all.

Last, even if population aging is very gradual, it is, nonetheless, inevitable, and the challenges posed by the emergence of "aging" societies will not be easier to tackle than those posed by rapid population growth. In view of the fact that all generations are concerned by this necessary societal adaptation, issues pertaining to the youth, adults, and the elderly must be addressed in tandem.

Conclusion: What Level of Population for What Level of Development?

The sweeping changes that are likely to affect population demographics in Sub-Saharan Africa will accompany socioeconomic transformations in an ongoing discussion:

- Population growth, an increase in demographic pressure, increased life expectancy, urbanization, and the aging of the population produce new economic and social behaviors among men and women, and among the young and the old.
- Conversely, social changes such as the evolution of the status of women, a change in mindset toward the age of marriage, the extension of educational opportunities, and, more generally, the changes in the organization of societies in terms of methods of production or attitudes, undoubtedly have an impact on demographics.

This adaptation by societies is undeniably more difficult today, owing to the speed and scope of the changes under way, particularly regarding globalization and a weakened role for the state. However, Sub-Saharan African populations have, throughout their history, demonstrated their capacity to adapt.

Demographic dynamics constitute one aspect of the transformation of societies that adapt on an ongoing basis to global development. This is clearly the case in Sub-Saharan Africa, where the three phenomena that were given particular focus (rural population growth, urbanization, and young age structure) are some of the challenges currently facing Africa. They should certainly not be viewed simply as problems but also as advantages to be seized. This is the challenge that must be met. And ongoing adaptation by Africa to this situation will be more effective if the populations, especially rural youth, are educated.

Annex: Necessary Population Policies

There is general consensus that a population policy is defined as a set of measures and programs that have been developed and implemented to have an impact on demographic variables (Gendreau and Vimard 1991). To this end, population policies are not only fertility control policies but also policies targeting the health and spatial redistribution of the population. In these areas actions may involve direct means (supply of information and services, persuasion, constraints) or indirect means (actions on the economic and sociocultural environment aimed at changing behaviors).

Like all policies, a population policy must, above all, be implemented to solve a problem or improve a situation. However, owing to the prevailing Malthusian mindset in our society, the characteristics of the populations in Sub-Saharan Africa are all too quickly perceived as problems that must be solved: the (too large) number of men (in fact, too many poor persons, too many potential migrants heading to cities or more developed countries), the (too high) population density (threat to the ecological balance), (too rapid) population growth, (too young) age structure, and the (explosive) urban concentration, because, despite the prudence displayed by scientists, "ideological discourse often uses the demographics argument to establish an assumed objectivity" (Véron 1993). Consequently, population policies are too often viewed solely as policies aimed at controlling fertility rates.

An Overdue Assessment of Population Issues

With the exception of a number of pioneering countries such as Botswana, Ghana, Kenya, and Mauritius, many governments in Sub-Saharan African countries did not have a specific population policy until recently. In efforts aimed at promoting development, certain measures such as improvements to the health system and rural development operations had a clear impact on demographic variables. Discussions on fertility and certain legal provisions (family allowances, maintenance of the French Law of 1920, and the like) were often pro-birth in nature.

It was after the initial world population conferences (Bucharest, 1974 and Mexico, 1984), and particularly after the Arusha Conference in Africa (1984), that the attitudes of governments evolved and there was a widely acknowledged need for population policies to be integrated into development policies—that is, population policies understood in the broad sense of the term based on the aforementioned definition and not only policies to control fertility.

All governments are committed to reducing mortality and promoting actions to achieve this objective, even if, as has already been seen, the current context is hardly conducive to a significant decline in mortality rates.

With respect to migration flows, it must be acknowledged that, more often than not, governments have only little room for initiative. International migration is spurred by the economic conditions in the host countries and the possibility of gaining access to these countries. The only actions considered are quite often ad hoc measures to expel foreigners.

Similarly, internal migration has generally been excluded from all planning; actions to colonize new rural areas are difficult to organize, costly, and target only small numbers, but they can serve as a trigger for steady migration flows. Attempts to stem the rural exodus or steer this group toward secondary urban centers have remained at the discussion level, except—yet again—with respect to ad hoc measures aimed at expelling urban dwellers toward villages, without any real effectiveness in the medium term and without respect for the rights of migrants.

Last, the principal donors have always paid particular attention to the fertility control issue and, on a number of occasions, exerted pressure on governments to encourage them to adopt birth control programs ("demographic conditionality"). As a result, discussions have evolved, although specific measures have not always been implemented. It bears noting that although a demand for contraception exists, especially in urban areas, the rural population (63 percent of the total population) is subject, more often than not, to conditions that are conducive to high fertility, as indicated in this chapter.

Policy Formulation and Implementation

In order to implement these actions, the approach followed over the past 20 years has often been the adoption of "a population policy declaration" establishing the guidelines and the framework for the actions to be carried out, followed by the finalization of "action programs" (Assogba 2003). In 2003 some 30 countries had adopted this declaration, while the process was under way in approximately 10. The earliest declarations have, in some cases, been updated (Mali in 1991 and 2003). They do not always give rise to concrete results and at times there are delays in implementing action programs (in Madagascar, the population policy was formulated in 1990, but the first program was adopted in 1997).

All these points reveal the difficulties encountered in designing and implementing population policies in Sub-Saharan Africa, especially given that structural adjustment programs imposed by the World Bank and the International Monetary Fund (IMF) were at considerable odds with any notion of a population policy: the reduction of social budgets (health, education) was responsible (at least partially) for stagnating, if not increasing, mortality rates and the decline in school enrollment (particularly among girls). And it is a well-known fact that these two points can but help perpetuate high fertility.

Fertility control, which cannot be decreed, can only be achieved gradually in the context of a general transformation of societies and their economic and social conditions. Fertility control policies undoubtedly require demographic "support," that is, "direct" policies (such as information campaigns, creation of a network of family planning centers, and contraceptive methods and services provided free or at a low cost). However, these policies cannot be implemented in isolation, without economic and social development policies, in other words, child health care, the promotion of women and greater equality in gender relations, literacy and education programs for mothers, the promotion of agricultural and nonagricultural activities, land tenure security, opening up of roads, access to water, and improved living standards.

More generally, serious obstacles, often stemming from fundamental issues, can only complicate population policies:

- Although African governments are now in a better position to consider population problems, the severe crisis affecting the continent is leading them to be less concerned with the medium and long term than with the short term (debt due dates; food imports; civil servant wage bills, when law and order are not affected; or maintaining their own grasp on power). This trend goes hand in hand with reduced planning (when it is not altogether absent) in a context that is heavily influenced by liberal ideology.
- Contrary to the expressed need to integrate population policies into development policies, government action today is divided into sectoral policies (population, health, poverty reduction, and so forth), which do not necessarily have an overall coherence. This division should be linked to the various "agendas" established at the international level (such as poverty, the environment, and the Millennium Development Goals).
- Despite the oft-expressed desire for democratization and decentralization, the relevant actors (the population, as well as intermediary bodies and nongovernmental organizations, play no part in the drafting of policies. The actual wishes of individuals and households are therefore not necessarily taken into account, and the autonomy of the social actors makes it possible for them to develop strategies (marriage, migration) that could render ineffective measures that have been adopted "by the upper echelons."

Population policies alone cannot solve development problems. A reduction in fertility rates is not the miracle solution that will enable Africa to emerge from underdevelopment and the crisis, for the roots of these problems originate from the complex interrelations between various phenomena that are undeniably related to population, but are also of a political, economic, and social nature.

Annex Bibliography

Assogba, L. 2003. *Population et développement en Afrique du sud du Sahara. National population policies. Basic concepts and tools.* UNFPA (United Nations Population Fund), Dakar.

Gendreau, F., and P. Viard, eds. 1991. "Politiques de population." *Politique africaine* 44. Karthala, Paris.

Véron, J. 1993. *Arithmétique de l'homme. La démographie entre science et politique.* Le Seuil, Paris.

Notes

1. The reader may check Ferry (2007) for a recent report on this topic.
2. To avoid having to revisit the issue, it bears noting at this juncture that, despite real progress, today there are still huge gaps in our knowledge of the demographics in Sub-Saharan Africa. In impoverished countries, which often have poor communication facilities, an unstable government, and no real statistical tradition, it should come as no surprise that population counts or the conduct of surveys are difficult exercises or that the civil service is inefficient. Moreover, demographic data have political significance and can be "sensitive" in certain countries (such as Gabon and Nigeria). Despite these difficulties, which exist in varying forms depending on the country, estimates of various demographic parameters that can at least be used for overall figures and for "significant" trends are now available.
3. Throughout the chapter, and unless otherwise indicated, "today" and "now" refer to mid-2005 for results pertaining to a given point in time, and to the 2000–05 period for those relating to a specific period. These results were provided by the United Nations for 2004 and 2005.
4. In this chapter, Sub-Saharan Africa includes Madagascar and the Indian Ocean islands, with the exception of the French territories of Réunion and Mayotte.
5. Cape Verde has long been a country of emigrants, with fewer of its citizens living at home than abroad, not only in the United States, but also in Europe (Portugal, France, Luxembourg, Netherlands, and Switzerland) and Africa (Senegal, Angola, São Tomé and Principe, and elsewhere).
6. In France fewer than 400,000 of its 4.3 million foreign-born residents were born in Sub-Saharan Africa (Institut National de la Statistique et des Etudes Economiques 1999 population census).
7. Since 1994 South Africa's economic power and prestige have made it a major host country for migrants, a growing number of whom are illegal.
8. As of January 1, 2006, there were over 13 million refugees around the world: the United Nations High Commissioner for Refugees (UNHCR) has assumed responsibility for 8.7 million people, and 4.4 million Palestinian refugees are under the care of the United Nations Relief and Works Agency. This figure does not include asylum seekers, repatriated, displaced, or stateless persons, or others who are also under UNHCR's care. All told, the UNHCR has responsibility for more than 21 million persons.
9. A number of recent studies have indicated that the impact of HIV-AIDS on these countries could, nonetheless, be less severe than expected.

10. Having numerous offspring appears to be related more to the absence of mechanization than to extensive production systems. A larger number of offspring has even been observed in intensive systems that require increased production per unit of land (in the Bamiléké and Mafa regions in Cameroon, for example).
11. Prevalence of contraceptive use is defined as the percentage of married women between the ages of 15 and 49 years currently using some form of modern contraception.
12. The United Nations' prospects include several "variants." We focus on the "medium variant."
13. The pace of the decline in the mortality rate is contingent upon three factors: medical advancement and the dissemination of these innovations to the population; the quality and effectiveness of the health system; and living conditions (such as nutrition, education, and income). This simple statement underscores the difficulty in formulating assumptions in this area.
14. Speech delivered by President Nicolas Sarkozy of France, at the University of Dakar, Senegal, on July 26, 2007.
15. The theory is also still being disputed by many African leaders who hold the view that people are more of an asset than a burden.
16. It bears noting that extensive farming in a manual economy continues to be more productive than an intensive development approach requiring a labor force that is often lacking when the agricultural season is short, except with respect to the use of labor-saving technologies (motorization, mechanization, chemical fertilizers).
17. In view of the challenges involved in the exercise of urban population projection, the United Nations cautiously prefers to adhere to this timeframe.
18. At this juncture we only briefly address the issue of the links between urbanization and development. These links, nevertheless, raise a number of questions, such as: Is a small job in the urban informal sector more beneficial to the nation than a job in the agricultural sector?
19. The issue of adaptation of land tenure systems with respect to population growth is well developed in two case studies conducted by CICRED (Committee for International Cooperation in National Research in Demography) on Burkina Faso (Drabo, Ilboudo, and Tallet 2003) and Niger (Guengant and Banoin 2003).
20. History may, in fact, help explain certain forms of settlement, for example, with appropriation of land by colonists in former colonies of settlement (Zimbabwe, South Africa, Namibia) or by the Church (Ethiopia).
21. Between 1995 and 2050, with a regular diet in place, vegetable food energy needs are expected to increase by more than five times in Sub-Saharan Africa, and even more than seven times in countries where staples include cassava, yam, taro, and plantain (Collomb 1999). The highly desirable improvement in the diet would increase this rate.
22. Moreover, the increase in international grain prices has paved the way for recapturing urban food markets that for a long time had been supplied by rice from South and Southeast Asia, and grains from Europe and North America. With dynamic agricultural policies, it is even possible that the trade balance of agricultural products could benefit not only from a decline in imports but also from increased exports.

Bibliography

Beauchemin, C. 2001. *Émergence de l'émigration urbaine en Côte d'lvoire. Radioscopie d'une enquête démographique (1988–1993)*. CEPED, Institut National de la Statistique de Côte d'lvoire, Paris.

Bloom, D. E., D. Canning, and J. Sevilla. 2002. *The Demographic Dividend. A New Perspective on the Economic Consequences of Population Change*, RAND Corporation. Santa Monica, CA.

Boserup, E. 1970. *Évolution agraire et pression démographique.* Flammarion, Paris. [English edition: *The Conditions of Agricultural Growth.* 1965. George Allen and Unwin Ltd., London.]

Bot, A. J., F. O. Nachtergaele, and A. Young. 2000. *Land Resource Potential and Constraints at Regional and Country Levels.* FAO, Rome.

Bryceson, D. F. 2004. "Agrarian Vista or Vortex: African Rural Livelihood Policies." *Review of African Political Economy* 31 (102): 617–29.

Collomb, P. 1999. *Une voie étroite pour la sécurité alimentaire d'ici à 2050.* FAO, Economica, Paris.

Drabo, I., F. Ilboudo, and B. Tallet. 2003. *Dynamique des populations, disponibilités en terres et adaptation des régimes fonciers. Le Burkina Faso, une étude de cas.* CICRED, FAO, Paris.

Ferry, B., ed. 2007. *L'Afrique face à ses défis démographiques, un avenir incertain.* APD-CEPED-Karthala, Paris.

Gendreau, E, C. Meillassoux, B. Schlemmer, and M. Verlet. 1991. *Les spectres de Malthus.* ORSTOM-EDI-CEPED, Paris.

Gubry, P., S. Lamlenn Bongsuiru, E. Ngwé, J. M. Tchégho, J. P. Timnou, and J. Véron. 1996. *Le retour au village. Une solution à la crise économique au Cameroun?* MINREST, IFORD, CEPED, L'Harmattan, Paris.

Guengant, J. P., and M. Banoin. 2003. *Dynamique des populations, disponibilités en terres et adaptation des régimes fonciers. Le cas du Niger.* CICRED, FAO, Paris.

Malthus, T. 1798. *An Essay on the Principle of Population. rev. ed.* 1980. INED, Paris.

Marchal, J.-Y. 1997. "Introduction. La ruralité ou la contradiction." *In La ruralité dans les pays du Sud à la fin du XXe siècle,* ed. J.-M. Gastellu and J.-Y. Marchal. ORSTOM, Colloques et Séminaires, Paris.

Marchal, J.-Y., and A. Quesnel. 1997. "Dans les vallées du Burkina Faso, l'installation de la mobilité." *In La ruralité dans les pays du Sud à la fin du XXe siècle, ed.* Gastellu and Marchal.

Mathieu, P. 1998. "Population, pauvreté, dégradation de l'environnement en Afrique: fatale attraction ou liaisons hasardeuses?" *Natures-Sciences-Sociétés* 6 (3): 27–34.

Quesnel, A. 2001. "Peuplement rural, dynamique agricole et régimes fonciers." *In Population et développement: les principaux enjeux cinq ans après la conférence du Cair,* ed. A. Lery and P. Vimard. CEPED, Paris.

Quesnel, A., K. Vignikin, B. Zanou, K. N'Guessan, and E. Vilquin. 1999. *Dynamique de peuplement des zones rurales libérées de l'onchocercose en Afrique de l'Ouest.* CICRED, FAO, Paris.

UNFPA (United Nations Population Fund). 2007. *State of World Population 2007. Unleashing the Potential of Urban Growth.* New York.

United Nations. 2004. *World Urbanization Prospects. The 2003 Revision.* New York.

———. 2005. *World Population Prospects. The 2004 Revision.* New York.

Véron, J. 1994. *Population et développement.* PUF, Paris.

Chapter 2

The Need for Inclusive Agricultural Growth at the Heart of Africa's Economic Transition

Bruno Losch

The rapid increase in its population has confronted Africa with the necessity of engaging in a strong process of sustainable economic growth, while relying on activities that are not very competitive in an international economic context that is highly so. Agriculture will remain the leading source of employment and income in Africa for decades to come, and thus it is vital to introduce government policies conducive to inclusive agricultural growth, drawing on the full range of market opportunities, while ensuring that the majority of producers are not left behind.

The weak performance of agriculture in Sub-Saharan Africa in both productivity and competitiveness poses a serious structural challenge for the continent, which must cope with its ongoing demographic transition and compete in an increasingly globalized and ever more competitive marketplace of agricultural and agrifood products.[1] This difficult situation is explained by the structural characteristics of African agriculture, its economic and institutional context, and the terms of its integration in international markets.

However, restricting analysis to the agricultural sector alone should be avoided. While the challenge to agriculture indeed exists in Africa, the overall economic context of the continent must be taken into account: some 50 years after gaining independence, the farming sector still weighs heavily in the economies of African states. This is shown by the overwhelming importance of agriculture to economic activity and employment, and also to gross domestic product (GDP) and trade. The overall context also underlines how weak the other economic alternatives are, especially the risks inherent in strong urbanization without industrialization, which does not create sufficient employment.

Sub-Saharan Africa is thus confronted with a major problem, namely, its capacity to absorb economically a rapidly growing population (1 billion additional inhabitants by 2050): in this context, given the perspectives of sector diversification in an extremely competitive international environment, agriculture will have to play a major role. The challenge facing African farming lies here, at the heart of the population–economic growth–employment nexus. African agriculture is obviously expected to contribute to feeding the continent and to fuel overall economic growth in a context of growing tension over food markets and natural resources), but it must also and will above all need to provide employment and foster activities that will directly or indirectly generate income for the huge number of young people entering the job market. The working population will continue to grow steadily, increasing by as much as 20 million additional workers a year by 2035, and African agriculture will have to accompany the continent's transition toward a more diversified economy able to progressively offer other employment alternatives.

The economic challenges facing African agriculture can be examined only in this overall structural context where demographics, the pressure on natural resources, and the international economic and political context all play their part in the process of change.

The debate regarding agriculture in Africa has long suffered from negative connotations because it has been considered synonymous with poor productivity, disinclined to change and weighed down by traditional structures, and destined to decline as cities expand. Modern approaches to industrialization adopted during the first two decades after independence, and the preference for agroindustrial-led development, progressively lost their steam because of failure or low returns on investment in a context in which international prices were anything but promising. Numerous attempts at integrated rural development undertaken in the 1970s and 1980s came up against difficulties regarding structural change, and especially the depletion of government resources, while donors focused first on macroeconomic stability and thereafter on across-the-board approaches aimed at poverty reduction. Despite this problematic situation, often made worse by policy bias in favor of the urban classes, some success stories were recorded in certain sectors or nationally, thanks to the dynamic nature of local stakeholders and their initiatives.

Increasing skepticism reigned around the year 2000, but now agriculture is belatedly recovering as a result of two main factors: recognition of its crucial role in reducing poverty, mainly to be found in rural areas; along with the instability of international agrifood markets, where recent overheating, after several decades of decline, poses once again the familiar question of how to "feed the planet." This change of attitude was recently made official by the publication of the 2008 edition of the World Bank annual report on development, devoted to the role of agriculture (World Bank 2007).

This document reports in depth on the current situation of world agriculture, its prospects for progress, and its role in reducing poverty. The report is important in that it proposes a customized approach to farming support, depending on the way structural change takes place in three different types of countries (the "three worlds of agriculture"): *agriculture-based countries* where agriculture still plays the major role; *urbanized countries* based on industry and services; and an intermediate stage where rural populations remain dominant but where agriculture plays a declining role (*transforming countries*).[2]

Over and above this welcome recognition of the role of agriculture per se, which nonetheless remains fragile because of increasing challenges worldwide, we feel that it is essential to place the problems facing African agriculture into a global context.

This chapter first shows that the farming question needs to be examined in connection with the structural challenges facing Sub-Saharan Africa. These challenges are somewhat unprecedented because of the "moment in time" and the way African economies do participate in globalization, and it is clear that agriculture is vital to the success of the growth-employment equation, in view of the ongoing demographic transition. The chapter then demonstrates that, in order to meet these structural challenges in a competitive international context where expectations are rising, government policies will have no choice but to facilitate inclusive agricultural growth that alone can offer income to the greatest number.

Agriculture and Economic Transition in Sub-Saharan Africa

By and large, the overall development agenda remains marked by the evolutionist model of structural change, as experienced historically during economic transitions—that is, the gradual transition from a primary structure to a tertiary one—in Europe.

According to this time-honored model, an increase in productivity of agriculture—the very first activity practiced by human societies (which explains why national accounts name agriculture as the primary sector)—allows capital to be amassed and the labor force to be released into other economic sectors, first manufacturing, then industry, and last services, which gradually predominate.

This process, which goes hand in hand with urbanization, does indeed reflect the transitions experienced initially in Western Europe, then in the United States, and by the rest of the so-called developed world. This has been corroborated by the economic transitions that have taken place over the past 50 years in a large number of developing countries in Latin America and in East, South, and Southeast Asia, and the countries that have made the most progress are now called, rather ambiguously, "emerging" countries (Gabas and Losch 2008; Coussy 2008).

This process of structural change, confirmed by the statistical evidence (Timmer 2007), should theoretically also take place in Sub-Saharan Africa. However, this vision is challenged by several analytical difficulties, which can be attributed to the angle adopted by the international thinking on development—an angle that results in bias in interpretation and anticipation.

The fact is that the major problem with the evolutionist approach, upon which development economics is founded, is its fixation on factor endowments and the economic, social, and institutional conditions of each country. This does not, of course, exclude the opportunities offered by comparative advantages in international trade, but it fails to reconnect internal processes with external processes and national trajectories reflecting historical specificities with the international context.

Because all things are not equal, the time when these processes of change occur matters, inasmuch as economic and political factors and the balance of power within the international arena at any given time determine the room for maneuver and affect the shape of development trajectories.

History and Current Specificities of Sub-Saharan Africa

Understanding globalization in its current forms—and, of course, the ways in which Africa copes with it—requires that these processes of change be examined in their global context (Gore 2003), and that a historical, self-centered comparison be set aside. Referring to transitions that took place in Europe can be useful in analyzing the mechanisms of economic transformation, but the example holds only for the world as it stood in the 19th and early 20th centuries. Indeed, two essential factors need to be underlined, since they were central to the agricultural and industrial revolutions in Europe. They are well documented, yet often minimized in the international debate regarding development and globalization.

First of all, from the 16th century onward, Europe fully benefited from its position of political and military domination, leading to colonial empires or spheres of influence.[3] This situation offered sources of supply and captive markets for Europe's burgeoning industry, protected by trade barriers at a time when any competition from outside Europe had been eliminated by territorial control or "unfair treaties" (Chang 2002).[4] This situation also allowed Europe to pursue industrial specialization while drawing its food supply from cheap imports (the British example was typical) guaranteed by the control of trade flows and privileged relationships with its colonies or dependent agricultural countries (Latin America).

Second, the transitions toward industry and urbanization associated with the agricultural revolution of the 18th century (Mazoyer and Roudart 1997) obviously took place in specific circumstances, reflecting the economic, political, and sanitary conditions of the times. They occurred over a period of some

150 years, accompanied by limited population growth, explained by high mortality caused by poor living conditions. They were later facilitated by large international migrations that played an important adjustment role (Hatton and Williamson 2005). Yet industrial growth, even though it remained firm, was not strong enough to absorb the labor force surplus resulting from rural depopulation.[5] Consequently, emigration, whether induced or spontaneous, appeared as a powerful exit option; about 60 million Europeans left for the "new worlds"[6] between 1850 and 1930, of which 35 million alone settled in the United States, a country that took in up to 1.3 million immigrants a year at the turn of the 19th century (Daniels 2003). These "white migrations" (Rygiel 2007) were possible only in the specific context of European hegemony.

Likewise, Asian and Latin American transitions need to be placed in context, since they are held up today as examples to demonstrate that this historical pathway toward structural change is inevitable. Situations are extremely varied among the countries of these two continents, but it should be recalled that these states were formed a long—or fairly long—time ago, and this proved essential for formulating autonomous development projects.[7]

These countries followed the self-centered policies characteristic of the international growth regime in force between the 1930s and the onset of the 1980s, when all countries—without exception—adopted national programs of economic consolidation based on domestic market development and protection, along with import substitution. This "turning inward," triggered by the 1929 crisis, lasted in various forms until the Fordist crisis of the 1970s, when the limits of the mass consumption model led to a clear slowdown of the main industrial economies.[8]

This is why many Latin American countries implemented industrialization-led modernization projects during the interwar years, adopted also by Asian countries starting in the 1950s. In addition, the Cold War between East and West played a nonnegligible role because a great deal of financial support from the United States served to consolidate certain governments. The result of this particular "moment in time" was the emergence of public policies targeted toward modernization, involving state intervention and considerable market protection, which, even though they were not always efficient economically speaking, did allow a fabric of companies, institutions, and a skilled labor force, which proved decisive during the ensuing period of international competition.

But the specificity of Sub-Saharan Africa in these matters must not be forgotten. First, the region was a latecomer on the international scene. Largely for geographical reasons, it long remained outside the world economy.[9] Then, when the continent finally opened up and during the following 150 years, European colonial powers clearly maintained it in a subordinate position, as a supplier of agricultural and mineral primary goods. Only in the decade preceding independence did these colonial powers agree, under local and international pressure, to set up the first programs introducing a proper economic and social infrastructure and laying the foundations of industrialization.[10]

Thereafter, independence was finally granted to African countries during the early 1960s, on the basis of former colonial territories, which had no relation, with some rare exceptions, to precolonial political structures. These young African states—with uncertain political structures, weak institutions, and very poor economic diversification—have been confronted since the 1980s with structural adjustment and especially with growing international competition, without having had the opportunity to implement lasting and true modernization policies.

All African countries did implement development programs. Many of them remained far from concrete; however, a number of industrialization projects were created, usually resulting in the neglect of agriculture.[11] Poor economic results and a growing and untenable debt, together with governance problems, have often been blamed for the failures of Sub-Saharan Africa. The above reminders show how important the historical processes and contexts are when comparing the stages of economic transformation.

A Novel Situation: Economic and Demographic Transitions in the Globalization Age

Another major feature—the demographic transition—has to be brought into the picture to highlight the integrated nature of the ongoing processes of change. A specific focus is needed because the short-term views of structural adjustment policies persistently neglected and ignored these long-term structural elements (Chataigner 2007).

This demographic transition, with a drop in the mortality rate before any comparable decline in the birth rate, results in very high population growth. Its pattern in Sub-Saharan Africa is characterized by a delayed takeoff, strong progress in health services, and a slow decrease of fertility (see chapter 1), explained by the slow pace of economic and social change. Thus, for the past 40 years, Sub-Saharan Africa has recorded an average annual population growth rate of over 2.5 percent, reaching 3 percent in the 1980s and only slightly decreasing thereafter. Above all, the high birth rate led to an extremely low activity ratio,[12] which fell below the one worker per one nonworker threshold during the 1980s and 1990s, which was also the time of structural adjustment. On the contrary, during this same period, East Asia, where the transition took place much earlier, fully benefited from its exceptional demographic situation, resulting in a ratio of 2.5 workers for every nonworker. South Asia, where the transition will take about 30 years, will not reap these same benefits (but slightly lower with a 2.1:1 ratio) until about 2035, and this difference can be put down to the radical birth control policy introduced in China. Sub-Saharan Africa will have to wait until at least 2050 to benefit from this type of favorable situation (figure 2.1).

Figure 2.1 **Activity Ratio, 1950–2050**

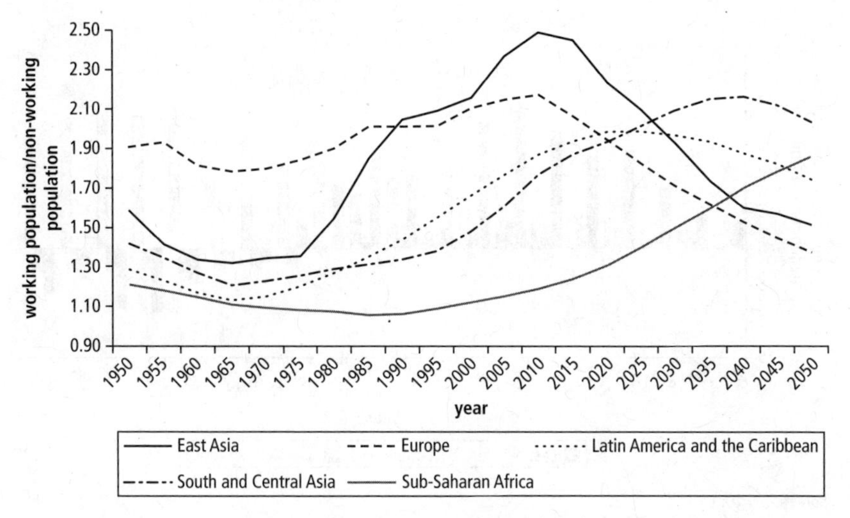

Source: UN 2006.

This long-term comparison sheds new light on the crisis that African economies have experienced and are still experiencing. Over and above the behavior that is so quickly held responsible for this crisis—rent seeking and bad management—a huge challenge was set that was difficult to achieve given the history of the young African states. But the challenge to come is even more daunting because the subcontinent will gain 1 billion inhabitants by 2050 and will have to ensure acceptable living standards for about 1.7 billion people.[13]

This situation has never occurred before, since these major economic and demographic structural changes will have to be handled simultaneously with globalization. This particular combination means that strategies for sustainable economic development must encompass population growth, in economies that have yet to diversify, in a highly competitive international economic context where product transportation, capital transfer, and information systems are being revolutionized, together with increasing economic openness leading to huge differences in productivity and competitiveness—all of which constitute a lasting handicap.

When the demographics of the continent are examined, not so much in terms of overall working and nonworking populations but rather in terms of job seekers, it is clear that African economies already need to absorb over 10 million new workers per year, a figure that will climb to roughly 18–20 million by the 2030s (figure 2.2).[14] For an average African country, with a current population of about 15 million, this means an annual total of 250,000 people, reaching 400,000 in 2025 (Losch 2006; Giordano and Losch 2007a, b).

Figure 2.2 Average Annual Growth of the Working Population, 1950–2050

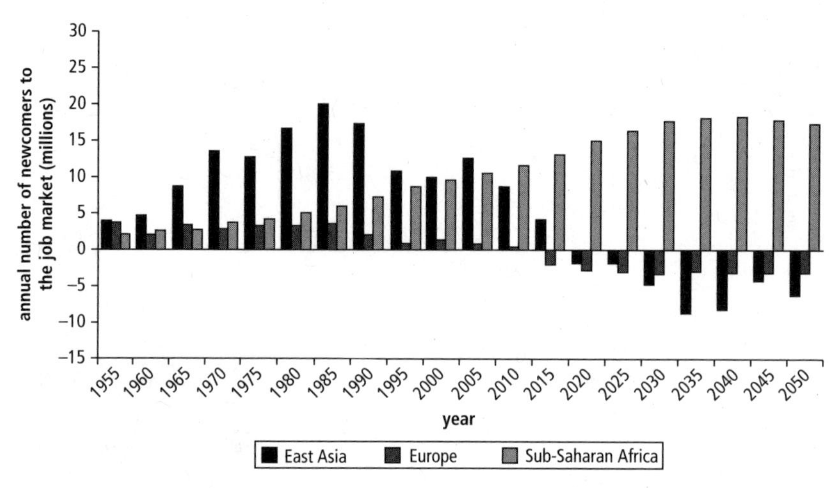

Source: UN 2006.

Demographic growth does, of course, signify an equivalent development in domestic markets, since the population needs to have access to goods and services, and this does and will continue to drive the economy—provided, of course, that the local economic, institutional, and political context is favorable and proves as efficient as that proposed by the international competition.

While the African population has doubled since 1980, the fact remains that a sustainable growth dynamic has not been set in motion. And while African agriculture has succeeded in keeping up with population growth (without any real gain in productivity and thus at the expense of a drain on natural resources), no real product diversification has occurred and no real urban economy has taken over (see below). This structural inertia has been unable to reverse a decline in per capita income since 1960.

The situation is such that, if there were no international migration barriers, a significant proportion of Africans stuck in a precarious state of survival would of course seek to emigrate, just like the poor European populations of the 19th and early 20th centuries, or even today, populations living on the borders of highly developed geopolitical zones (Mexico and Central America neighboring the United States, or Morocco and Turkey neighboring the European Union).

But the capacity for economic, political, and cultural absorption into a "finite world" has its limits (Valery 1945), at least until the working-nonworking population ratios of certain developed countries (or China) have dropped.[15] The poor populations of Sub-Saharan Africa seem to be lastingly trapped in their native lands and the fact that they have no means of escape leads Pritchett

(2006) to say, somewhat provokingly, that many African countries could turn into nations of zombies.

African Agriculture, Vital to the Growth-Employment Equation

To avoid being bogged down in these pessimistic forecasts—even if they are not unrealistic—we need to return to the fundamental aspects of African economies to sketch out some possible fields of action.

As pointed out by the *World Development Report* on agriculture (World Bank 2007), countries in Sub-Saharan Africa constitute the archetype of agriculture-based economies. Urbanization over the past 50 years has certainly left its mark, since the urban population has multiplied by 10—from 25 million to 235 million (or from 13 to 35 percent) between 1960 and 2004[16]—but these countries remain deeply rural and dependent on farming.

According to the Food and Agriculture Organization, the agricultural population of the region stood at 415 million in 2004, or 62 percent of the total population.[17] In many countries (25 out of 47, South Africa excluded), the farming population is even higher than the rural population, for two main reasons: the existence of a dynamic urban and periurban agriculture, especially in the large metropolitan centers (a phenomenon not specific to Sub-Saharan Africa), and the fact that residents of many small or medium-size towns maintain some farming activity nearby, which is typical of African urbanization.[18]

Thus, the main characteristic of African economies is the overwhelming role of agriculture, which still occupies on average almost 65 percent of the working population (the median is closer to 70 percent in the 47 countries of Sub-Saharan Africa). Although a number of recent analyses state the contrary,[19] the activity structure has not changed a great deal over the past half century: the active population in agriculture only decreased by 25 percent on average—20 percent in median terms (29 countries remain below this 20 percent of change). The only real changes, apart from specific cases that are hardly representative (small oil-producing countries or island states), can be observed in Nigeria and slightly less in Côte d'Ivoire and Cameroon.[20] On the other hand, in Asia and Latin America, economic structures have changed much more rapidly (often registering a decrease of 60 to 70 percent).[21]

Agriculture remains vital for generating value and in trade structure. For the average African country, the contribution of agriculture to GDP has remained at around 30 percent since the early 1980s, with lower figures for oil- or mineral-producing countries and higher figures for countries involved in, or recovering from, war. Industrial activity mainly centers on the mining sector, while the budding manufacturing sector has stagnated for the past two decades with some rare exceptions (Mozambique, Uganda, Zambia). The service sector consists mainly of trade and informal services (government employment has been reduced), requiring few technical skills (such as financial intermediation or technical backup). Trade is of course closely connected with farming. As for exports, they have long demonstrated that Sub-Saharan African countries contribute to the international economy through primary activities, particularly mineral and farm products.

In view of these inherited shortcomings, as well as weak economic and institutional fabric and low human capital, African economies have little room for maneuver in terms of diversification. The productivity gap compared with the rest of the world is a huge handicap for international competitiveness and, since Africa has been unable to gain market share by increasing production, it will long remain marginalized with respect to world trade.[22] In the world market for farm products and food, Africa accounted for 10 percent of exports in 1960; by 2005 this figure had dropped to under 4 percent.

However, African economies are highly integrated into the global economy in terms of trade. Foreign trade (imports and exports of goods and services) weighs roughly 45–50 percent of total GDP, that is, a higher level than that of the OECD. This general remark only goes to show how weak domestic African markets really are, since they rely on very low average incomes.

A tenacious trend has taken root, postulating that urban growth itself fuels development (Cour 2007). By creating a market for agriculture, urban growth is supposed to encourage both the growth and modernization of agriculture, and activity diversification is supposed to improve overall productivity as a result of increased industrial returns. In the wake of a tenfold increase in the urban population of Sub-Saharan Africa over a 40-year span, the subcontinent's structure and economic performance hardly reflect the existence of a virtuous circle of this sort.

Here again, the "moment in time" and the historical context need to be included in the debate. The urbanization processes of the 19th and much of the 20th centuries were inherent to the structural changes that occurred in a given place at a given time, as described above (periods of European domination followed by self-centered growth). Progress went hand in hand with slower demographic transition, but it also coincided with industrial development, leading to increased productivity.

Urban development over the past quarter-century has been quite different. Populations have grown more rapidly than economic development, leading to an extraordinary spurt of the informal sector, which was promoted for several decades and did act as a cushion, but which is also a low-productivity sector[23] with underemployment, lack of job security, and low returns—all of which culminate in the creation of urban slums, which are springing up all over the world (UN-Habitat 2003; Davis 2006). This phenomenon is, of course, exacerbated in Africa because of the structural problems experienced throughout the continent, marked by the way it initially entered the global economy and by the limited scope and degree of autonomy of its economic policies. Consequently, African countries encounter difficulties in developing new activities in this global competitive context. The lack of urban employment structuring the economy was reported by Todaro (1971), using the term "underemployment traps," and the 1980–90 world crisis made the situation even worse.

Urbanization in Africa is urbanization without industrialization; but for towns to stimulate both agricultural markets and technical progress, urban demand has to be creditworthy, and not just based on minimal demand and nutrition needs. Demand even has to be the driving force, instigated by urban activities providing increasing returns.

Need for Inclusive Government Policies

African Economic Transition and Room for Maneuver

To avoid political and social upheaval, the main economic challenge for Africa today is to cope with a rapidly growing population and the numerous young people joining the job market each year. The major objective is to develop industries and services able to absorb this growing labor force and to contribute to improving overall productivity. The experience of the past 40 years has shown that government decrees will not make this transition happen, since, apart from a few limited opportunities, the major shortcoming is lack of competitiveness, caused by weak economic, institutional, and legal contexts, along with quality issues and production costs.[24] The needed transition is more likely to result from a long-term approach that sets policy priorities.

Today, given the structural inertia that has been prevalent on the continent since independence (its specialization in primary activity and its employment structure as described above), its low-productivity urban development, and the problems it faces regarding international competition made worse by globalization, it is vital for agriculture to carry on absorbing labor supply. In order for the economic transition to take place smoothly, agriculture will probably need to fulfill this role for the next 30 years while the population continues to grow, together with other sectors where employment needs to be developed, for example sectors reliant on population and urban growth (construction, transportation, education, health, and the like).

The proposals put forward by the *World Development Report 2008* (World Bank 2007) for eliminating rural poverty—which is crucial to achieving the Millennium Development Goals—are useful in offering a general framework for analysis as well as a useful reference. According to the report, three strategies exist for eliminating rural poverty: increased agricultural specialization and market integration, especially in international markets; diversification through the development of nonagricultural rural activities and growth in wage labor (farming and nonfarming); and migration, toward cities or abroad. Specialization and the development of farming enterprises obviously entail adaptation to markets' requirements, which are increasingly demand-driven with new quality standards, meaning that technical and financial skills must be available. Activity diversification toward trade, crafts, or services—via self-employment or the labor market—remains an option for rural dwellers who do not possess the financial, human, or social capital necessary for agricultural specialization. Migration remains an alternative for those unable to find employment locally.

This three-pronged framework, which reflects the dominant evolutionist view, can be applied in most cases to address the problem of rural poverty. But it is not necessarily suitable for Africa—in fact, the report's recommendations are less systematic for agricultural-based countries —and its viability derives from the relative importance of each option within the possible combinations. This is because the adaptation challenge differs according to whether the combination is 40/40/20 (40 percent farm specialization, 40 percent rural diversification, 20 percent migration) or 10/30/60. The report makes quite clear that these options are not unique. History has shown that transitions are always multiform and often combine all three options (farming and nonfarming activities and income and migratory flows). It should nevertheless be noted that this combination is currently becoming more systematic, since there are few unique alternatives allowing for a definitive change of activity or place of residence. So which options remain open?

The migration option raises the question, generally avoided in the literature, of how it can actually be put into effect. Which urban activities are sustainable despite the internal and external obstacles hampering their development and are also capable of ensuring massive and lasting employment? Which regions remain open to international migration? There is a great deal of migration within the subcontinent, and African countries are the main destination of migrants (Ratha and Shaw 2007). But these migrations are unduly emphasized. Not enough attention is given to the absorption capacity of destination countries (the example of Côte d'Ivoire is particularly telling), and the same economic and social realities are shared in the end with other countries in the subregion. Except in the event of large-scale territorial and political remapping, these migrations are restricted by existing national boundaries. As for population movements toward northern countries, these are clearly limited by current geopolitical forces and the migratory policies adopted by these countries (that is, European selected immigration).

Rural activity diversification has historically been an integral part of the transition process. In developed countries, this diversification has gone hand in hand with increased farm productivity, rural depopulation, and industrialization. Literature abounds on the subject of economic development and has, since the founding work of Lewis (1954) and later Johnston and Mellor (1961), emphasized the crucial role of links,[25] because the development of nonagricultural rural activities is closely connected with growth in the farming sector itself, as well as an increase in the level of qualifications. Growth in agricultural production creates a demand for inputs and stimulates processing, transportation, and marketing activities; also, increased income for farming families leads to increased consumption of goods and services. The experience of Asian countries clearly shows this transition to be important, because it fosters and consolidates activity diversification and urban development.

But these observations bring us back to square one. First and foremost, it is agriculture that drives the transition process, and agricultural development is more important than specialization per se, which leads back to the overall context surrounding agricultural growth and especially the opportunities and constraints inherent in the different types of market.

Opportunities and Constraints of Agricultural Markets

Development opportunities offered by agricultural markets need to be analyzed in connection with the way African agriculture is organized. This remains largely based on family farms, which have small production units and overwhelmingly employ family labor; access to land is often not secure and working capital is limited, especially in terms of equipment, which restricts the area that can be farmed per worker.[26] All of these aspects hinder development, which also suffers from difficult access to markets and high risks associated with an economic and institutional context both unstable and damaged by the 1980–90 economic crisis (Bélières et al. 2002; Bosc and Losch 2002). All this helps to explain why self-consumption and partial product marketing continue to exist.

Some family farms blessed with greater assets (land, financial resources) are able to directly undertake agricultural diversification with a view to supplying markets, using paid workers, and thus engaging in economic differentiation. This process was long hindered by an unfavorable economic context, including major direct and indirect taxation and scarcity of public goods, which limited returns. Further setbacks were incurred when agricultural extension systems were closed down after state withdrawal which, in the absence of a sufficiently consistent and viable demand, did not encourage private initiatives. The consequence is incomplete markets (particularly regarding the supply of inputs such as seeds and fertilizer, extension and advisory services, credit, and risk management) but also imperfect markets, incurring high transaction costs for matching supply and demand (accessibility, information, moral hazard), costs that were previously covered by government marketing systems for certain sectors. These systems were often costly, but the risks connected with product sales were lower.

These farming structures are now confronted with changes in the global economic context, clearly providing new opportunities but also new constraints, and both have significantly modified the structural transformation process.

The first factor of change involves the restructuring process that began in global agrifood markets in the 1980s and translated into growing economic integration. This integration stems from the increased involvement of private stakeholders after governments withdrew their control of markets, as well as to privatization. This was intensified by takeovers involving large firms in the food sector and also by the openness that goes hand in hand with globalization (decreased tariffs, direct foreign investment).

This integration is, generally speaking, a demand-driven process, governed by new consumer needs, involving a demand for higher quality resulting from improved living standards and concomitant changes in diet. It covers both vertical integration through high-value chains, managed by processing firms, and the use of new distribution networks, commonly called the "supermarket revolution" (Reardon and Timmer 2007). The integration has changed the rules of the game, deeply affecting marketing conditions and the ways farm products are sold and purchased. Whereas demand was previously uniform, product differentiation now exists, based on new quality and traceability requirements and the development of norms and standards, often accompanied by the development of contracts.

These changes offer great market opportunities, but the new requirements also call for adaptation involving technical expertise, financial means, and access to information networks. Consequently, these changes become major discriminating factors that can foster integration or exclusion for family farmers.

The second factor of change, often unexplored by research, is the direct result of what could be called the "confrontation effect" (Losch 2006) between completely different levels of productivity and competitiveness, created by this openness to international markets. It is possible today to produce and to consume on a global level, with a growing disconnection between places of production and places of consumption, facilitated by the development of transportation and communication (products, information, capital movements). This sea change puts different types of agriculture with widely different levels of productivity in competition, in addition to market distortions created by the subsidies granted by numerous developed countries (the United States, countries of the European Union, Japan). These productivity gaps reach at least 1:1,000 for cereals (Mazoyer 2001) when manual farming with low inputs—as in Sub-Saharan Africa—is compared with the highly capital-intensive farming practiced in several parts of the world (Western Europe, North America, the Southern Cone of Latin America, Australia), hence significantly increasing the risk of marginalization for less efficient types of farming and farmers.

Competitiveness is not just a question of costs (the most frequently quoted aspect) and quality of products (increasingly complex and demanding), but it also requires "volume," which is essential for increasing market shares (the undisputable indicator of competitiveness). But volume of supply is also a direct result of productivity, and, in this new context where agricultural markets are increasingly interlinked, the risk is that the more productive farm systems will capture the growing global demand—directly related to the increasing world population and the new quality requirements stemming from diet changes (Collomb 1999).

This new context, together with the characteristics of African agriculture, allows three major types of markets to be distinguished, reflecting different requirements and room for maneuver for African farmers.

The first market type is probably the best known and has often been presented over the past few years as the way of the future, given that global prices were generally on the decline: high value-added products, that is, horticultural products including fruits, vegetables, and flowers, directly linked with new global distribution systems, but also niche products for specific market segments (organic and fair trade products). Horticultural products have overtaken traditional tropical commodities in terms of export value, but market specificities require high quality and major investments in water control, packaging, and storage, all of which pose technical and financial barriers to market entry for small farms. On the other hand, given these requirements, such products need a large labor force and constitute a source of employment in the form of agricultural wage labor.

The second market type covers tropical commodities (coffee, cocoa, rubber, vegetable oils), which have made a considerable contribution to putting Sub-Saharan Africa on the international economy map, and where the continent enjoys several enviable market positions. These products are more easily handled by small family farms, but international prices were long unfavorable and local prices much affected by high transaction costs. Competitiveness in these markets is clearly determined by economies of scale and the level of supply, which lead back to productivity requirements.

The third and final market type is the food commodity market destined primarily for local and regional consumption. Farmers naturally put their surpluses on the market, and accessibility using local networks is high. Potential for development is driven by population and urban growth, and quality requirements are less demanding than for exports (animal products differ from crop products). Competition from low-priced imports (for certain meat products and cereals) was formerly a constraint, but pressure has now diminished as a result of the sharp rise in global food prices; in addition, a degree of natural protection exists for land-locked countries (and remote regions), along with a degree of "cultural protection" connected with dietary habits, consumption patterns, and food preparation specificities (that is. the importance of root crops and plantains).

Which Priorities for African Agriculture?

Given the variety of opportunities and constraints presented by agricultural markets, a number of options can be envisaged.

Advantages currently held in traditional export markets should be maintained and consolidated, because family farmers have easy access to them and income can be widely spread. The challenge is to increase productivity and improve marketing channels in order to diminish costs, offer better returns to producers, improve quality, and gain market share.

On the other hand, high value-added products are difficult for small farmers to develop because they require economic agents able to make the necessary investments and cope with the marketing logistics, and who agree to place at least some portion of their source of supply with local farmers. The trend currently observed in most countries is the gradual marginalization of small farmers, who, in some cases, had initially been involved in developing these sectors, as they are replaced by specialized firms that directly control production or medium-size farms under contract with a view to diminishing transaction costs (Huang and Reardon 2008). Investment opportunities in these sectors do, however, need to be grasped and facilitated because they create jobs and generate income, they can create packaging or processing infrastructures, and they promote activity diversification in rural areas.

But the main objective must remain uppermost. Horticultural exports can contribute significantly to growth in the farming sector and to the trade balance, but in absolute values they contribute little to improving income in rural areas. At best, and depending on the country, a few tens of thousands of jobs have been created directly or indirectly, while the future of hundreds of thousands or even millions of farmers is at stake.[27]

Given that the main problem facing African economies is the creation of jobs and income-generating activities, the priority for government policy makers is to improve the situation for the greatest number and to help family farms by supporting the development of food products and markets. These markets represent about 70 percent of the total value of African farm products (exports and locally consumed products combined), that is, about $50 billion a year (Diao et al. 2007). Development of food markets is essential because Sub-Saharan Africa is currently running a food deficit, and another major incentive is the significant rise in import prices.

Food markets are quite clearly the most inclusive markets. They are open to all and are essential for earning income, potentially reducing poverty and linking up with other sectors thanks to increased consumption. Improving the performance of local, national, and subregional food markets will diminish the risk for farmers and thereafter enable them to invest and also to diversify as the need for self-supply goes down. Last, production growth can stimulate the development of processing activities, create local value-added, facilitate diversification, and intensify rural activities directly connected with the local urban fabric.

In this context, priority needs to be given to two strategic tasks: an increase in productivity,[28] and improved performance and access to markets, all being vital if production, income, activities, and jobs are to progress. Production increases and improved market supply will of course lead to a gradual drop in the price of farm products, which is good news for consumers and labor costs, but farmers can be more than compensated by higher product volume resulting from greater yields from the land and higher output per worker.

The issues that should be tackled by government policy makers are discussed in later chapters (see part 2), but a few guiding principles should be recalled.

Direct intervention affecting the prices of farm products should of course be avoided, because costly and counterproductive market distortion occurs; efforts should be directed instead to improving production itself. The chief priority is, without a doubt, to develop the provision of public goods, which are sadly lacking throughout Sub-Saharan Africa: research, particularly on improved varieties[29] and farming techniques suited to the climatic and economic conditions, transportation and infrastructure (irrigation and electrification grid), data and statistical systems, training, and land rights. These efforts will help to solve the vast majority of market imperfections and related transaction costs and reduce risks.

At the same time, there is a real need to solve the difficult problem of "incomplete" markets, penalized if not crippled by missing links attributable to the absence of private agents. Such is the case in the seed and fertilizer market, essential for improving production but also for coping with increasingly poor fertility. Such is also the case for farmer backup in terms of advice and extension services. And last, the most difficult problem of all concerns financial services and risk management.

For these different fields of action, government support is usually required, in various ways and for various periods. This support can also take the form of collective action and backup for farmer organizations or professional bodies, since their ability to make proposals and act upon them has long been marginalized as a result of the "political expropriation" of rural areas, whose weight in the democratic and election processes falls well below their demographic weight (any comparison with the influence of agricultural lobbies in urbanized and industrialized countries reveals a glaring discrepancy).

It should be recalled that the priority given to food production was the mainstay in the structural transformation of Asian economies, with the clear objective of managing and slowing the exit from agriculture and at the same time diminishing food costs; this approach had a marked impact on alleviating poverty (but less of an impact on reducing disparities). These objectives were reached thanks to considerable gains in productivity, which allowed production and farm income to develop, and to linkage mechanisms, which were all-important to the development of nonagricultural rural activities. Irrigation, seeds, and fertilizer were the main ingredients in this green revolution, made successful by massive government investments in infrastructure, research, and extension, and also by subsidies provided for fertilizer, together with price protection and support that were the hallmark of the self-centered policies of the 1960s and 1970s.

Today, international markets and economic policies are constrained by increased openness. But if transition is to succeed over the next few decades, African agriculture and African economies as a whole will need to make a massive effort regarding investment and support and will need to work toward greater regional integration.

Thus, action in favor of agriculture remains vital. Agriculture lies at the heart of the growth-employment nexus and will be required to absorb employment needs for the next two or three decades until activity ratios change and annual growth in the working-age population starts to decrease. The current pressure on international prices for agricultural products creates an auspicious moment in time and an opportunity for reinvestment that should not be missed.

This action requires the close attention of government decision makers, as well as that of the international community, particularly donors. It is a question of the balance between agriculture, rural depopulation, and urban growth, and how these have been managed historically, because this is the key to development. It has always been a major preoccupation of governments dealing with the delicate issue of the distribution of activities and population within their territory (Coussy 2006). Faced with the risks of overly rapid urbanization and excessive rural depopulation, governments have always adopted policies that try to reconcile the internal and external opportunities and constraints of each period of history. During the 19th century, many European countries encouraged international migration, while others implemented protectionist agricultural policies. During the 20th century, East Asian and South Asian countries fully supported agriculture, thus coping with transition by increasing income and creating solvent demand. Today, China is still trying to reconcile industrialization and urbanization with forceful control of internal migration.

Sub-Saharan Africa is the only exception to this trend of close attention to the balance between demographics and the economy, mainly because of the historical difficulties encountered in the emergence of African states on the international scene. For governments and development agencies, it is high time to address the African challenge by repositioning agriculture at the heart of the continent's economic transition, based on policy choices that are conducive to inclusive growth and that offer income and work opportunities to the many.

Notes

1. Some of the points and examples developed in this chapter are drawn from an expertise and research program—the RuralStruc program (*Structural Dimensions of Liberalization in Agriculture and Rural Development*)—set up by the French government (Ministry of Foreign and European Affairs, Ministry of Agriculture and Fisheries, Agence Française de Développement, and Centre de Coopération Internationale en Recherche agronomique pour le Développement), the World Bank, and the International Fund for Agricultural Development. All the opinions expressed here are personal to the author and do not necessarily reflect those of the World Bank Group or of the other institutions involved in this program.
2. Seventy-four developing countries with a population above 5 million are ranked on the basis of two criteria: contribution made by agriculture to growth, and the ratio of rural poverty to total poverty. Taking this rural poverty ratio and the selected data sources, Senegal, the Democratic Republic of Congo, and Zimbabwe are considered to be countries in transition and no longer agricultural countries.
3. The conditions that led to European hegemony cannot be developed here, but the discovery and domination of the Americas (Grataloup 2007)—with their own specific history—appear to have been crucial events.
4. Bairoch (1997) points out that in 1750, India and China accounted for slightly over half of world manufacturing production. In terms of unfair treaties, the best-known ones are those that were imposed by European powers and the United States on China and Japan.
5. This powerful process toward growing cities, motivated such as it is today by poverty and exclusion (resulting in particular from increased agricultural productivity), was facilitated by the development of the railways.
6. The estimates vary, depending on whether migrants who returned to their home countries are counted. The main destinations were the United States and Canada, Argentina and Brazil, Australia and New Zealand, South Africa, Algeria, and also Siberia. These were "new worlds" only from the European perspective, with European civilization giving little credence to indigenous peoples.
7. Irrespective of their individual national histories, the Latin American states have existed since the beginning of the 19th century. The state systems of many Asian countries are directly based on previous experience, and those that are former European colonies usually obtained independence between 1945 and 1955. The situation in Sub-Saharan Africa is quite different in this respect.
8. As later events show, this would prove to be the main stimulus behind the new period of globalization, since companies had to seek growth in competitive external markets, because their national markets were no longer sufficient engines (Giraud 1996). Both governments and international institutions followed suit, resulting in the current "consensus."
9. According to Grataloup (2007), this was attributable mainly to the barrier constituted by the Sahara, hostile climatic conditions, and the difficulty in penetrating the continent given the dearth of means of transportation (mainly sea and river) until recent years.
10. This was the "period of the planners" (Hopkins 1973) and of the British Colonial Development and Welfare Acts and the French FIDES, implemented in particular

by the Caisse Centrale de la France d'Outre-mer, precursor to the Agence Française de Développement.

11. The rent-seeking-based theory developed by the new political economy school, notably by Bates (1981) and the Berg report (World Bank 1981) in Sub-Saharan Africa, widely used to justify state withdrawal from markets, made much of the urban bias manifested by the new elites to explain why agriculture was neglected. This explanation is only partially true (Losch 2000).
12. Usually, the dependency ratio is given, that is, the ratio of the nonworking population to the working population. Since our point of view examines activity and employment, we use the activity ratio (workers to nonworkers), which is more explicit.
13. Recall that, contrary to economic forecasts, demographic forecasts are calculated using actual existing populations, and that approximations simply reflect different hypotheses regarding birth and mortality trends that are relatively stable. The only significant variations are caused by natural catastrophes, health hazards, or the consequences of political events.
14. In view of the incomplete data on job markets in Sub-Saharan Africa and the extent of underemployment, the 10 million figure reflects the number of newcomers to the job market seeking one or several sources of income, and thus economic independence. These numbers are equal to the annual growth of the working-age population (15–64 years of age, according to the international standard).
15. In 2020 China will have to start coping with a drop in the working population and a sharp drop in the rate of activity, which is likely to lead in 2050 to the same structural problems encountered in Europe, but with a population of 1.4 billion (and 330 million over 65 years of age).
16. The annual growth rate of towns was high at first (almost 5 percent), then dropped considerably from the 1980s onward (slightly under 4 percent) in reaction to the economic crisis.
17. Agricultural population is defined here as the total number of people whose livelihood depends on agriculture (used in the widest sense of the term), that is, the working population involved in farming and their dependents.
18. Certain statistical biases should be noted. Generally speaking, the definition of "urban" means a dense cluster of people (often 2,500 inhabitants), but in some countries, the definition is purely administrative. This is the case in Senegal, for example, where localities that have few urban attributes have been officially designated as *communes* and are considered to be urban districts. Consequently, almost 50 percent of the population is classified as urban in a country with a farming population estimated at 70 percent.
19. The *World Development Report* on agriculture (World Bank 2007) mentions a "considerable" decrease in the region's farming population (p. 27), taking as an example Nigeria, which is significant because of its size (being the most populous country, with a population of 130 million) but atypical and nonrepresentative of the other countries (see below).
20. Cameroon and, above all, Nigeria, are well-known examples of the "Dutch disease." The case of Côte d'Ivoire should be considered apart because of the overall situation in the country (and the way it records statistics).

21. The particular cases of China and India should, however, be noted, since this indicator should theoretically place them in the same group as the African countries. Here, a low rate of structural change reflects neither absolute values nor scale, since several hundred million workers have left the farming sector for jobs in industry, thereby contributing to industrial development, which has led to high overall growth. But these two countries are, nonetheless, responsible for the future of 500 million and 300 million people, respectively, who still work in farming (along with their dependents).
22. Overall productivity is calculated by applying value-added to total working population. The gap is about 1:5 between Africa and other developing countries, and 1:100 when compared with developed countries (roughly $500, $2,500, and $50,000 respectively per worker, in constant values based on the period beginning in 2000). UNCTAD 2006.
23. Ranis and Stewart (1999) distinguish between two informal subsectors: a traditional subsector of the so-called "sponge type" stemming from the surplus of agricultural labor, with incomes sometimes lower than rural incomes; and an informal subsector now undergoing modernization that revolves around the formal urban sector.
24. With the exception of labor. But labor costs are not a sufficient advantage and comparisons need to encompass technical skills and equivalent qualifications. Sub-Saharan Africa is then at a disadvantage because of the low level of professional training, worsened by the ongoing crisis in the education system.
25. Among many sources, the reader may refer to Johnston and Kilby (1975), Timmer (1988), and, more recently for Sub-Saharan Africa, to Delgado, Hopkins, and Kelly (1998).
26. Family agriculture, typified by a family-based structure of activity in terms of residence, assets, means of production, labor, and decision making (as opposed to the model of managerial farms), is the main way in which agriculture is organized worldwide. The gaps between different types of family farming revolve around productivity and income (see below).
27. This is particularly true in Kenya, which has become the leading exporter of cut flowers to the European Union and where horticulture is now the second most important export industry. There is, however, no precise information on the employment generated by this sector (as is often the case): in Kenya, where the horticultural "success story" of the continent has taken place, about 40,000 farmers are reportedly involved out of a total of 5 million farms.
28. In view of the critical importance of job creation, this objective should be given careful attention so as to avoid inappropriate technical choices (meaning, for example, that the initial emphasis should perhaps be on animal traction and light, rather than heavy, motorization).
29. Compared with Asia, food commodities are more varied (root crops, plantains, so-called secondary cereals) and have generally been less addressed by international research.

Bibliography

Bairoch P. 1997. *Victoires et déboires. Histoire économique et sociale du monde, du XVe siècle à nos jours.* Gallimard, Paris.

Bates, R. H. 1981. *Markets and States in Tropical Africa. The Political Basis of Agricultural Policies.* University of California Press, Berkeley, CA.

Bélières J.-F., P.-M. Bosc, G. Faure, S. Fournier, and B. Losch. 2002. "What Future for West Africa's Family Farms in a World Market Economy?" Issue Paper 113, IIED, London.

Bosc P.-M., and B. Losch. 2002. "Les agricultures familiales africaines face à la mondialisation. Le défi d'une autre transition." *Oléagineux corps gras lipides* (4): 450–65.

Chang, H.-J. 2002. *Kicking Away the Ladder: Development Strategy in Historical Perspective.* Anthem Press, London.

Chataigner J.-M. 2007. "Avant-propos." In *L'Afrique face à ces défis démographiques. Un avenir incertain,* B. Ferry, ed. AFD-CEPED, Karthala, Paris.

Collomb, P. 1999. *Une voie étroite pour la sécurité alimentaire d'ici à 2050.* FAO, Rome; Economica, Paris.

Cour, J.-M. 2007. "Peuplement, urbanisation et développement rural en Afrique subsaharienne: un cadre d'analyse démo-économique et spatial." *Afrique contemporaine* 223: 3–4.

Coussy, J. 2006. *Les questions démo-économiques.* Working paper, RuralStruc program.

———. 2008. "Emerging Countries: An Attempt at Typology." In *Emerging States: The Wellspring of a New World Order,* Ch. Jaffrelot, ed. C. Hurst & Co., London.

Daniels, R. 2003. *Guarding the Golden Door: American Immigration Policy and Immigrants since 1882.* Hill and Wang, New York.

Davis, M. 2006. *The Planet of Slums.* Verso, New York, London.

Delgado, C., J. Hopkins, and V. Kelly. 1998. *Agricultural Growth Linkages in Sub-Saharan Africa.* Research Report 107, IFPRI, Washington, DC.

Diao, X., P. Hazell, D. Resnick and J. Thurlow. 2007. *The Role of Agriculture in Development: Implications for Sub-Saharan Africa.* Research Report 153, IFPRI, Washington, DC.

Gabas, J.-J., and B. Losch. 2008. "Fabrication and Illusions of Emergence." In *Emerging States: The Wellspring of a New World Order,* Ch. Jaffrelot, ed. C. Hurst & Co., London.

Giordano, T., and B. Losch. 2007a. "Structural Change in Agriculture: Confronting the Transition Issue." 45th Annual Conference of the Agricultural Economics Association of South Africa, Johannesburg, September 26–28.

———. 2007b. "Transition: Risques d'impasse?" *Courrier de la planète* 81–82: 22–26.

Giraud, J.-N. 1996. *L'inégalité du monde. Economie du monde contemporain.* Gallimard, Paris.

Gore, C. 2003. *Globalization, the International Poverty Trap and Chronic Poverty in the Least Developed Countries.* CPRC Working Paper 30, UNCTAD, Geneva.

Grataloup, C. 2007. *Géohistoire de la mondialisation. Le temps long du monde.* A. Colin, Paris.

Hatton, T. J., and J. G. Williamson. 2005. *Global Migration and the World Economy: Two Centuries of Policy and Performance.* MIT Press, Cambridge, MA.

Hopkins, A. G. 1973. *An Economic History of West Africa.* Longman, London.

Huang, J., and T. Reardon. 2008. *Keys to Inclusion of Small-scale Producers in Dynamic Markets: Patterns in and Determinants and Effects of Farmers' Marketing Strategies in Developing Countries.* Synthesis Report-Micro Study, Regoverning Markets. IIED, London.

Johnston, B. F., and P. Kilby. 1975. *Agriculture and Structural Transformation: Economic Strategies in Late-Developing Countries.* Oxford University Press, New York.

Johnston, B. F., and J. W. Mellor. 1961. "The Role of Agriculture in Economic Development." *American Economic Review* 51 (4): 566–93.

Lewis, W. A. 1954. "Economic Development with Unlimited Supplies of Labor." *Manchester School of Economics and Social Studies* 22 (2): 139–91.

Losch, B. 2000. "Eloge de la distinction. S'intéresser aux intrigues pour comprendre les situations africaines." *Economies et sociétés* 34 (8): 87–114.

——— 2006. "Les limites des discussions internationales sur la libéralisation de l'agriculture: les oublis du débat et les oubliés de l'histoire." *OCL* 13 (4): 272–77.

Mazoyer, M. 2001. *Protéger la paysannerie pauvre dans un contexte de mondialisation.* FAO, Rome.

Mazoyer, M., and L. Roudart. 1997. *Histoire des agricultures du monde.* Le Seuil, Paris.

Pritchett, L. 2006. *Let Their People Come: Breaking the Gridlock on Global Labor Mobility.* Center for Global Development, Washington, DC.

Ranis, G., and F. Stewart. 1999. "V-Goods and the Role of the Urban Informal Sector in Development." *Economic Development and Cultural Change* 47 (2): 259–88.

Ratha, D., and W. Shaw. 2007. *South-South Migration and Remittances.* World Bank, Development Prospect Group, Washington, DC.

Reardon, Th., and C. P. Timmer. 2007. "Transformation of Markets for Agricultural Output in Developing Countries since 1950: How Has Thinking Changed?" In *Handbook of Agricultural Economics* (vol. 3): *Agricultural Development: Farmers, Farm Production and Farm Markets,* R. E. Evenson and P. Pingali, eds, 2808–55. Elsevier Press, Amsterdam.

Rygiel, P. 2007. *Le temps des migrations blanches. Migrer en Occident (1850–1950).* Aux Lieux d'être, Paris.

Timmer, P. 1988. "The Agricultural Transformation." In *Handbook of Development Economics,* vol. 1, H. Chenery and T. N. Srinivasan, eds, 275–331. North-Holland, Amsterdam.

———. 2007. *A World Without Agriculture: The Structural Transformation in Historical Perspective.* Wendt Lecture, American Enterprise Institute, Washington, DC.

Todaro, M. P. 1971. "Income Expectations, Rural-Urban Expectations and Employment in Africa." *International Labour Review* 104 (5): 387–413.

UN (United Nations). 2006. *World Population Prospects: The 2006 Revision.* Geneva.

UNCTAD (United Nations Conference on Trade and Development). 2006. *Développer les capacités productives.* Report on the least developed countries. Geneva.

UN-Habitat. 2003. *The Challenge of the Slums: Global Report on Human Settlements 2003.* Earthscan Publications, London.

Valery, P. 1945. *Regards sur le monde actuel et autres essays.* Folio, Gallimard, Paris. 1988.

World Bank. 1981. *Le développement accéléré en Afrique au sud du Sahara. Programme indicatif d'action.* Washington, DC.

———. 2007. *World Development Report 2008: Agriculture for Development.* Washington, DC.

Chapter 3

Ecosystems: Reconciling Conservation, Production, and Sustainable Management

Jean-Jacques Goussard and Raymond Labrousse

The significant production potential of Africa's biomass cannot conceal its vulnerability: resources to be shared in increasingly contentious situations, careful management of the exploitation of woody flora and the replenishment of forest reserves, soil erosion and the fragility of living landscapes in the wake of deforestation, and management of water resources in the context of climate change. Options related to the capacity of ecosystems to reconcile food or energy production with the conservation or reproduction of ecological capital should first be discussed with local stakeholders.

In its recent *World Development Report 2008*, the World Bank (2007) highlights the paramount importance of agriculture in Sub-Saharan Africa for achieving the Millennium Development Goal of reducing by half the proportion of people living in extreme poverty and suffering from chronic hunger. The importance of the sustainable management of natural resources is underscored in this report, thus marking a departure from the tendency to accord a secondary role to natural capital in strategies aimed at combating poverty and inequality.

With a largely rural population, the goods and services produced by the natural ecosystems of Sub-Saharan Africa provide a vital foundation for all development processes—particularly agricultural development—by compensating, to some extent, for the fairly limited investment capacities in general. However, all the specificities of these ecosystems need to be taken into account to understand why such vast and often sparsely populated tropical spaces cannot better feed an ever-growing African population.

This chapter provides insight into governance trends in natural resource management, followed by a description of the different Sub-Saharan ecosystems. It then undertakes a discussion of how the management of these ecosystems should be positioned between conservation and production, and then linked to the various environmental and agricultural challenges that need to be considered.

Natural Resource Governance: A History of Trials

The natural layout of African landscapes has been greatly modified and reshaped by the impact of human activity, which has taken different forms in recent centuries. Clearly, demographic growth remains a key factor in these changes (see chapter 1). The pace of change in natural resource management increased significantly during the colonial era, displacing traditional practices that were based on

- Customary systems for the use of space, with specific decision-making channels and spiritual relationships between individuals and the various components of ecosystems
- Conservation of strategic natural resources (natural capital) through traditional methods of organization, which, though sometimes complex, were also specific (code for the use of pastureland during the dry season in the interior delta of the Niger; sanctions to encourage protection of the *Acacia albida* in the Sahelian zone, and so forth)

Contrary to the colonial vision of an open space "without assets or masters," given that land registers were not used, all ecosystems were covered by land rights or user rights, even pastureland used for short periods in arid zones or forests described as "virgin" (see chapter 6, which addresses land-related problems). Long-standing land rights, such as planting or cutting down trees (right of clearance) are still in place today. However, in some instances, custom-based land rights proved inadequate to protect natural environments, as was the case with chainsaw registration, which affected a significant number of Ivorian forests cleared during the 1970s.

The colonial era was characterized by the introduction of land laws patterned after rights observed in Europe. In some instances certain resources and land were seized in the territories based on these new rights, while in others there was an acceptance of vaguely or better-known custom-based land rights.

Conservation responsibility for natural resources was quickly transferred to the colonial administration. Forest services were established and organized based on a paramilitary model, with territories being extensively partitioned on the basis of land and rural codes. The formidable precision of these codes was matched only by their unsuitability to the contexts in which they were applied (for example, these codes required authorization to cut down each tree that had long been owned by farmers or villages).

A network of protected spaces (reserved forests, national parks, forest reserves) was gradually established through "legal procedures" that were resented by local populations witnessing the transfer of their custom-based rights to the state. This intrusion by the state into the deepest rural areas marked a break with tradition; spaces that were once "unbroken," appropriated, negotiated, lived in, and farmed were suddenly fragmented, zoned, and separated by profound discontinuities. The effects of this period persist to this day, as particularly evidenced by the following:

- The affected populations, feeling disenfranchised, viewed protected areas as land to be repossessed.
- The preservation of swaths of land for purposes of natural ecosystems in rural areas that were largely degraded, stripped of their original features, and made artificial came at a high cost, taking into account their value (nonwoody products, pharmaceutical products, environmental services, and the like) to rural populations and their role in biodiversity conservation and the regeneration of natural systems in cases where land can still lie fallow for long periods.

In many African countries, particularly francophone countries, the postindependence period was characterized by an orientation toward a planned economy and the preeminence of the state, which sought to control the management of open spaces. The advocates of development assistance believed in the unlimited opportunities offered by controlling nature through the application of technical solutions deemed to be optimal. Change was supposed to come through cognizance by an enlightened population of the benefits of progress. Agricultural and forest production projects increased in number, with consideration being given in some cases to complementary social, but not environmental, components.

The failure of many ambitious projects aimed at domesticating nature and the questions raised regarding the consequences of the droughts of the 1970s and 1980s played a role in undermining certainties. The debates that started during this period are still taking place today. They include the respective roles of imported techniques and endogenous innovations as well as the effectiveness of sectoral polices versus locally developed approaches. However, with the increase in the number of land management and natural resource projects, a deeper knowledge of the ecosystems and methods integrating environmental considerations also developed over time. These projects faced an array of obstacles such as inappropriate legislation and inadequate consideration of farmers' economic strategies.

Attempts to review the legal underpinnings and prerogatives of ecosystem management in the context of rewriting rural, forest, and land codes were rarely successful. Too often these efforts entailed minor revisions that were almost cosmetic and sidestepped the issues of the legitimacy of custom-based rights and the consideration of regional specificities.

In some countries, land management concerns were considered in the context of local development and decentralization frameworks in order to address deficiencies in the area of territorialization. The transfer to local governments of a number of natural resource management prerogatives, particularly in the forestry area, met with resistance from administrations that were more concerned with their loss of power than with the benefits of playing a new role of advising populations and their representatives.

At the national level, the establishment of protected areas—an issue actively promoted by international organizations since the 1990s in the wake of the Rio Conference—has often contributed to heightened competition with farmers and herders for the use of space. The forestry (and often "fauna") administration has trouble keeping up with rapid change in a context where the roles and utility of natural ecosystems are gradually being recognized, rediscovered, and expanded well beyond forest production only. National environmental plans are being prepared at the same time as ministries bearing the same name are being created. The latter, which often do not have deconcentrated services, have difficulty fitting into an institutional framework that, for the most part, has been in place since the independence era. The ascendancy of civil society and advocacy groups are, however, helping shape a new political landscape in an area where recognition of the importance of this issue is gradually increasing.

Pending the widespread adoption of new instruments such as payment for environmental services, local participatory approaches can provide a space to take account of the political need to link the Rio commitments with the central issue of poverty reduction.

African Landscapes

Based on an initial and broad overview, a distinction can be made between East Africa, Southern Africa, and Africa west of the Congo-Nile ridge. In the western part, the ecoclimatic gradient typically seen reflects the diversity of natural ecosystems, from the dry Saharan and Sahelian zones to the more humid equatorial regions. In East Africa, the major landforms generally shape the diversity of landscapes depending on altitudes. The natural systems of Africa are generally formed around a few vast river valleys (Niger and interior Niger Delta, Nile valleys, Congolese basin).

Dry Desert or Subdesert Zones

In the Saharan zone, where rainfall is extremely low and unpredictable, vegetation is limited to very short-season flora and a few rare hardy and shrubby species. Subdesert zones, where rainfall is slightly higher (up to 200 millimeters), allow for the development of a steppe that includes a few hardy gramineaceous and salsolaceous species, short-season annuals, and a few shrubs and trees (acacias). Production systems are limited to nomadic livestock activity and oasis agrosystems. Consequently, in this region, agriculture is merely one component of a system of multiple activities based largely on commerce and trade. Population densities are obviously very low. Desertification and the impact of major droughts on herds have greatly affected livestock activity, leading to the rapid and ever-increasing migration of populations to urban centers. More and more, traditional activities are under assault by new realities such as increasing motorized activity in deserts, new opportunities, discovery tourism, and mining activities.

The resurgence of latent conflicts in the Saharan zone—the consequence of carving up land during the colonial era regardless of ecological or cultural factors—along with age-old antagonisms among nomadic peoples or with their southern neighbors, and even the radicalization of conflicts linked to terrorism or religious fundamentalism, are, unfortunately, possible scenarios, against a backdrop of growing tension over scarce resources. Last, geopolitical uncertainties are thwarting more effective interstate antilocust efforts.

The decline in nomadic livestock activity, already under pressure, could intensify as a result of climate change and more severe droughts, quickening the pace of migration to urban centers on one hand and producing greater sedentarization as a result of concentration in oases on the other.

The Sahel: The Shifting Agricultural Frontier

With rainfall fluctuating between 200 and 400 millimeters per year, vegetation in the northern Sahel is composed of steppes, pseudosteppes, and savannas dominated by annual grasses with a high capacity for postdrought regrowth. The tree stratum is dominated by acacias and bushy species. This ecological mosaic is directly dependent on pedological conditions and the water retention capacity of the soil. These dry regions also shelter vast systems of continental humid zones (interior Niger Delta, Lake Chad, and the Okavango River in Southern Africa).

Production systems are still often dominated by livestock activity organized around local and/or widespread transhumance over long distances toward the south during the dry season (see box 3.1 on the pastoral economy).

Rainfed agriculture is random at the local level; in some places it has been abandoned while in others it is being reintroduced, depending on whether meteorological cycles are more humid or drier. All of this makes for an agricultural frontier that is dynamic and fluid. Under these conditions, harnessing water resources takes on a special importance, whether for purposes of irrigated agriculture or pasture irrigation.

BOX 3.1

What Does the Future Hold for the Pastoral Economy?

The situation in the livestock zones of the Sahel is unpredictable from a climatic standpoint and often unstable from a political point of view (Julien 2006). The herders have developed a survival economy focused on the movement of humans and herds. In Chad and Niger, contrary to the policy of "modernizing" sedentarization recommended by a number of donors, the Agence Française de Développement has supported a series of projects aimed at increasing and providing safe conditions for the movement of livestock and at strengthening this activity by enhancing the traditional method of extensive exploitation through the provision of water for grazing areas and the management of pastureland around water points.

The establishment of new water points with low water flows in order to avoid overgrazing facilitates the establishment of new grazing areas and slows the movement of migratory animals toward the predominantly agricultural southern zones, while the creation of migratory network routes, which were mapped out by herders and farmers, helps reduce conflicts with farmers.

Consequently, in the context of the current conflict in eastern Chad, an increase is being seen in animal production and the preservation or even enhancement of pastoral natural resources. External assistance supports natural resource management by herders and farmers in a context of dialogue and organized participation, with the assistance of specialized parties (such as nongovernmental organizations and consulting firms). Initiatives of this nature require provisions for heavy investment in human capital.

Source: Jullien 2006.

Permanent agricultural activity is found more in the south (annual rainfall between 400 and 800 millimeters). The woody flora is richer there and rural landscapes are enriched by new plant communities, such as lowlands and gallery forests. There are few short-season cultivars suited to these conditions (millet, sorghum, fonio, and cowpea), and given the shorter rainy season, farmers everywhere are looking for shorter-season plant varieties. These often come from more arid zones, such as sorghum from Mali. Pastoral resources are shared in increasingly antagonistic conditions, with transhumance livestock activity from the north clashing with fairly sedentary crop and livestock activity. While rainfall conditions are more favorable in this zone, the profitability of inputs used is highly dependent on rainfall levels.

Irrigated agriculture can be carried out using natural means, through connection to the permanent or temporary rise in waterways or through controlled means, using boreholes or irrigated areas. This agriculture is clearly of strategic importance in the Sahel.

The rapid degradation or even disappearance of humid zones that can be used for irrigated agriculture connected to high water areas is a source of concern for the functioning of vitally important natural ecosystems (for example, winter season areas for palearctic migrants). The future of large swaths of irrigated areas is very much dependent on the efforts made to maintain them, to manage irrigation water, and to remove drainage water. Furthermore, the harmonization and coordination of water and basin management policies are dependent on interstate initiatives to manage the major river basins (Niger Basin Authority and Organization for the Development of the Senegal River). In broader terms, the stakes related to sustainable irrigated agriculture go well beyond agricultural production issues. Important issues need to be resolved concerning water development (for example, new techniques such as drip irrigation) and balances need to be struck in the management of the natural ecosystems of humid zones.

The development of market gardening, particularly in the dry season, often in urban or periurban areas, is an essential source of food for populations, for diversifying the activities of men and women, and for controlling migration toward cities.

Obtaining wood (for energy, services, and woodwork) remains a major problem in regions where local production is limited owing to ecoclimatic constraints, and where demand is exploding in areas such as those close to major urban centers (Dakar, Bamako, Niamey, Ouagadougou, and N'Djamena). Sustainable solutions should also be found to control the exploitation of woody flora and manage wood resources. This process should be started by involving the populations concerned, using models like the wood market village management model in Niger.

In many areas, the chances of continuing transhumance livestock activity in a context of increasingly dense population of rural spaces and greater expansion of protected areas are becoming increasingly remote, particularly at the southern boundaries of the Sahelian region (transition zone of the Sudan), where transhumance areas are gradually being converted into agricultural areas.

Last, it should be noted that the situation facing major fauna in West Africa is generally critical, compared with East Africa where special attention has long been paid to the fauna, given its economic importance (tourism and hunting activities). The issue is, therefore, one of striking a future balance between agriculture and livestock activity, wooded areas and crop-growing areas, and protected areas and areas exploited to meet the needs of the population.

A Long, Dry Season in the Sudanese and Sudano-Guinean Zones

In this extensive Sudanese zone, which stretches west to east from Senegal to the Central African Republic and where rainfall is fairly significant (between 800 and 1,800 millimeters), the main constraint is the duration of the dry season, which currently lasts for five or six months. However, from north to south, there are three different zones:

- Sudanese (800 to 1,200 millimeters), with the hardy, annual graminaceous flora of the savanna, diversified, fire-resistant woody flora of varying densities, and narrow, discontinuous gallery forests surrounding the humid lowlands.
- Sudano-Guinean (1,200 to 1,800 millimeters), predominated by hardy graminaceous flora, tree vegetation composed of Sudanese species and, locally, of humid forest. The mosaic distribution pattern of these forest systems and fairly dense tree patches are essentially the result of the soil type.
- Transition to the forest zone: this singular area is found in the upper savanna belts (Guinea, Central African Republic, and Democratic Republic of Congo) in contact with the humid forests. It is characterized by the dearth of fire-resistant woody plants and the great fragility of the remaining forest relics (which, once cleared, will be difficult to regenerate as an original forest ecosystem).

This vast Sudanese zone is characterized by the variety of transitions, from the bushy savanna to robust graminaceous formations and dry forests. The effect of fires is important and ever present, affecting at times even the fringes of lowlands and large gallery forests that are still intact and house expanses of flora similar to those found in the humid forest. Expanding areas of fragile duricrust soil stripped of tree flora can be found in this zone. Some dry forests of Southern Africa (Angola, southern Democratic Republic of Congo, Zambia, western-central part of Madagascar) have profiles that are somewhat similar to those of the dry forests of the Sudan zone.

One section of the Sudanese zone is still quite sparsely populated compared with the overall area (20 to 60 inhabitants per square kilometer). However, pressure on cultivated agrosystems is often high, given the surface area that is sufficiently fertile to grow annual crops.

In the most sparsely populated areas (under 20 to 25 inhabitants per square kilometer), family-based agriculture is often relatively widespread, using a slash-and-burn approach for woody flora while selectively sparing a number of species considered useful (such as the shea and nere trees). Based on this system, land fertility is restored though woody fallow (five to fifteen years depending on the environment), with diversified biotic regrowth and a significant buildup of organic material linked to the root system of tall grasses. Short-season food crops (cowpea, maize, millet, peanuts, and sorghum) are dominant and frequently alternated with cotton cultivation. These practices are increasingly under threat as greater pressure is exerted on land, often by the arrival of migrants in north Cameroon or southwest Burkina Faso, for example). Fallow periods are becoming shorter and permanent fields are being established in the most fertile areas.

The risk of physical soil erosion is high, owing to the torrential rains and the protracted dry season, particularly in instances where the soil covering the duricrust (frequently found in West Africa) is not very dense. This degradation is exacerbated by the development of animal traction farming and the clearing of wood.

Rearing of ruminants in this context combines various components: migration from the Sahel, seminomadic herders, farmers increasing their herds, and speculative livestock activity by urban dwellers. The practice is still very widespread and is often based on the use of fire to stimulate the regrowth of pastureland in savannas with hardy grasses. Controlling these fires and the more destructive late-season fires, which are often caused by traditional hunting, remains a significant challenge.

Hunting activities remain significant in the most sparsely populated zones. In some instances, these activities are led by traditional hunting brotherhoods of well-organized groups that may sometimes assume responsibility for managing the fauna just outside protected areas (experience with the Upper Niger National Park in Guinea). Forests are often conserved or even extended by prohibiting access to grazing animals (village agroforests in Upper Guinea).

Higher population density and demographic pressure (from humans and livestock) are shortening fallows and the size of pastureland. Use of crop residue and dung manure remains insufficient to compensate for lower soil fertility. Aging fruit groves (*Acacia albida*, shea, nere) face other regeneration threats, which are increasing because of the rapid reduction of tree fallows and the clearing of land to make way for animal traction farming. Tree and wood densities in savannas (which largely determine the goods and services yielded by these ecosystems) are dependent on water availability and the level of disturbances (fires, pastureland, and the like). Climate change could drastically affect these equilibriums, with the herbaceous strata being the most vulnerable (Sankaran et al. 2005). The gradual disappearance of swaths of forest and the depletion of sacred woods are seriously undermining the chances of regenerating forest areas where fires have been prohibited, owing to the growing scarcity of forest tree seeds. Local management of protected areas (protected forests and reserves) still needs to be improved. While increased land pressure is affecting vast areas of southern Mali and Burkina Faso, northern Nigeria, and northern and central Cameroon, occupation of other rural areas remains moderate (southeastern Chad, Central African Republic, and onchocerciasis areas), a situation that points to possible flexibility in the context of effective space planning and migration.

Future opportunities lie, first and foremost, in the remarkable biomass production capacity of these ecosystems.

The Intertropical Humid Zone

Postforest Areas. In the intertropical humid zone, forests have for the most part disappeared in many countries, from West Africa to Central Africa: the Guinean coastline, part of southern Sierra Leone, Liberia, Côte d'Ivoire, Ghana, Nigeria, São Tomé and Principe, central Cameroon, the southwestern part of the Central African Republic, a portion of north Angola, the lower Congo, the coast of north Mozambique, and northwest Madagascar. Mention should also be made of the special case of the four-season, low rainfall, humid zones in south Benin and Togo, the Congolese Niari, western Democratic Republic of Congo, and northwest Angola. These areas all have one or, more often, two short dry seasons lasting less than three months, along with ecosystems that were initially composed of dense humid forests and fairly extensive mangrove areas along the coasts.

Traditional agriculture is based on the slash-and-burn method and food production of mainly root vegetables, tubers and, locally, rain-fed rice. Land rights are traditionally based on the clearance process. In these systems, crops and fallows are alternated for periods of varying lengths, depending on occupation density. A few useful trees, such as the palm oil tree, are preserved during clearance activities. Forest ecosystems have been almost completely degraded outside of protected areas, forests reserves, and protected forests subject to some degree of conservation.

This rapid degradation occurred as a result of pressure exerted by converging factors, such as the development of major plantations (rubber from Liberia) or a variety of industrial, often farm-based crops (coffee and cocoa), forest activity, and invasion by the *Chromolaena* (a Lao shrub), which prevents any kind of forest regeneration. The situation has worsened with migration from the north (need for labor for plantations, the Sahelian drought, and land needs) and the "race" (involving all actors) to establish vast estates, quite often with very partial agricultural use of this land.

In light of this degradation of forest ecosystems, national strategies aimed at restoring forest resources based on crop planting (in Côte d'Ivoire, for example) have been promoted, most often using exotic species that grow fairly quickly and are suited to the market. Clearly, these planted crops alone do not yield the same level of environmental goods and services as the natural forests they have replaced.

In postforest areas, the expansion of protected areas is limited. These areas are under constant pressure, a situation that clouds prospects for conservation, particularly of the major fauna. This is the case with protected areas, in the strict sense of the term, and also with protected forests, specifically forest relics that still exist in rural areas. In these areas, sacred forests are also disappearing completely.

Sustainable plantation agriculture faces many challenges, among which is the restoration of the aging capital from coffee and cocoa cultivation and the forest soil capital that once supported their development but that is currently degraded by the unfortunate practice of full-sun plantings. The return to bigger shaded systems compatible with forest production (*framiré/fraké*) remains a current solution that allows for adaptation to the many soil, climate, and erosion constraints as well as to phytosanitary risks.

In these intertropical humid zones, a significant area of progress lies in the expansion of small-scale agroforestry that facilitates diversified and sustainable food production. In the case of short-season agriculture, undersowing often seems well suited to agroclimatic conditions, a situation that may even allow for mechanized undersowing agriculture similar to the kind used in Brazil.

Urban expansion is also opening up markets for horticulture and short-term livestock activities (Asian model), provided that the health risks associated with the latter are well managed. In these areas where conditions are, for the most part, manmade, production of timber and construction wood will probably follow private sector plantation development (in particular, teak).

In coastal areas, the future expansion of shrimp farming, which will very likely take place over time, will require a major effort in spatial planning and site management. This process should also provide for the conservation of the mangroves, which are necessary for the functioning of coastal systems, maintaining fisheries' diversity and richness, and for fish farming itself. It should be noted that the expansion of coastal cities and the pollution associated with this phenomenon are reflected in the irreversible changes in lagoon and coastal environments (the alarming factor is the magnitude of these changes, given that the purification capacity of these areas has often been greatly surpassed). In continental waters, improved fish farming pilot exercises (Cameroon, Côte d'Ivoire, upper Guinea) are expected to lead to organized industries.

In postforest areas, one of the main issues is how subsistence farming and agroenterprises will evolve to take advantage of expanding urban markets and opportunities to export such tropical products as coffee, cocoa, pineapple, rubber, and palm oil.

Sparsely Populated Dense Forest Areas. Such forests are located mainly in the big Congolese basin and adjacent regions (southeast Cameroon, Central African Republic, Gabon, continental Equatorial Guinea, Republic of Congo and the Democratic Republic of Congo). Historically, these regions have been populated by successive waves of migrants from the north, a situation that has progressively marginalized the indigenous hunters and gatherers (pygmies).

Climate conditions are characterized by rainfall distribution, either with or without a short dry season, conducive to the growth of dense and diverse forest stands, depending on the number of species and vegetation phases (structure).

In these forests, roots are concentrated less than half a meter from the soil cover, the biotic renewal process is slow, and restored fertility is concentrated in the first 30 centimeters of the soil, hence the very great fragility of the soil once the forest disappears. The regeneration of numerous plant species calls for fauna requiring very different conditions. Numerous forest species require shade when plants are young. Other than the land forest, mention should be made of the modest amounts of alluvial forests (Congolese basin) and the stands in humid zones—palm swamps and marshes. Also, savanna formations resulting from both fire and soil characteristics (Gabon, Congolese plateau) are limited, because the forest flora of neighboring ecosystems do not have fire-resistant species, a factor that limits the possibility of the regeneration of forest flora.

Knowledge of these complex ecosystems is very limited, and the control of the ecological impact of forestry activity—even that qualified as sustainable—does not measure up to stated ambitions. This does not apply to a number of valued, light-dependent species such as the *okoumé* or palm oil trees. In the strict sense of the term, deforestation in annual values is, however, fairly limited (0.4 percent). In the Congo basin, a sustainable management approach has been adopted since the late 1990s with respect to major industrial concessions, as a result of the efforts of states, the private sector, research, and donors, primarily France. Of a total 55 million hectares of forest concessions granted, a sustainable management approach is applied to 60 percent. These management plans are extending conservation efforts to some degree by fostering:

- An exploitation approach that takes into account the renewal of forest resources
- Maintenance of the social role of forests (sacred forests)
- Stabilization of watersheds and waterway systems by maintaining a better-preserved forest cover rather than applying an agroindustrial or slash-and-burn approach to agriculture. Maintenance of the forest cover helps mitigate the impact on biodiversity (composition, structure, and organization of populations), which has nonetheless not been well evaluated.

Owing to low levels of demographic pressure in dense forest zones compared with other parts of Africa, the need for agricultural land along with long fallows was, for a long time, in keeping with existing needs, even though the periodic movement of agricultural villages had significantly impacted forest space dynamics for centuries. This situation has changed in recent decades, particularly as a result of the gradual degradation of forest ecosystems linked to forest clearance and logging. Populations have for a long time settled close to roads, either because of various pressures (consolidation of population groups during the colonial era) or simply to have easier access to services. In countries with substantial oil revenue, these populations have migrated toward urban centers.

With respect to agriculture, except for a brief period of aggressive action to improve agriculture in the former Belgian Congo (quickly thwarted by a labor shortage), most efforts to develop agriculture and a plantation economy have failed, even when heavily subsidized. Health constraints linked to cattle rearing (trypanosomiasis) have prevented the establishment of an extensive pastoral system based on the conversion of forests, as in the case in Amazonia.

The issue of the future of forest ecosystems in Central Africa raises the following questions:

- How can biodiversity conservation imperatives (not only in terms of specific wealth but more particularly the integrity of the functioning and structure of these natural systems) be reconciled with the acceptable and sustainable economic development of forest resources? Greater understanding of these ecosystems is clearly necessary, as is ensuring compliance with sustainable management commitments, which must be linked in a practical manner to conservation efforts.
- How can the effects of hunting and the wild meat trade, often linked to forest activity, be controlled, given that these activities often pose a grave threat to the integrity of natural ecosystems?
- How can the potential risk of massive forest clearance for agricultural purposes, patterned after the disastrous Amazonian pioneer front, be avoided? The initiative on Reducing Emissions from Deforestation and Forest Degradation in Developing Countries (REDD), adopted in Bali in December 2007, could be helpful in this regard. It could also help prevent the rapid exploitation of forest resources driven by the needs of the Asian market and its rapid development. The most isolated forests, accessible as a result of large-scale road or rail development projects involving the Congo basin, would, therefore, be threatened. The financial aid provided by those countries most interested in accessing resources (the announcement by China regarding the doubling of aid and its concentration on infrastructure) points to the likelihood of these trends.

The outlook for industrial agriculture in these regions is more hypothetical, largely because of soil fragility, but also because large areas are unsuited to mechanized agriculture. Labor is also in short supply and the lack of access to roads is widespread.

High-Altitude Humid Forests. Forests, briar, and pseudosteppe formations above 2,500 meters once constituted the main landforms in Cameroon, Kenya, Ethiopia, the Nile-Congo crest (Rwanda, Democratic Republic of Congo, Burundi, and Uganda), and the area stretching from Malawi to South Africa. Currently, these natural formations are found only as somewhat isolated relics in densely populated rural areas that sometimes exceed 500 inhabitants per square kilometer. Climate and health conditions are conducive to settlement and account for the sharp increase in inhabitants and fairly extensive cattle rearing. These favorable conditions have led to a very diversified agriculture, combining tropical crops (tea and coffee) and temperate crops, allowing two short crop seasons a year.

The very significant decline in natural conditions seen today is linked to the overcrowding of land and the complete privatization of space. The reduction of natural forest spaces poses real risks to the original plant and animal communities and to a number of typical species, such as the mountain gorilla.

The main challenge remains the great pressure exerted on land in many areas. Hence, there is the risk of continued conflict and difficulties to improve the standard of living of farmers who are trying to live off very small plots of agricultural land, which often measure less than 1,000 square meters per person in a household.

Natural Capital: Constraints and Opportunities

African ecosystems harbor important natural capital. Sound use and development of soil, water, flora, and fauna can facilitate not only the conservation but also the enhancement of this capital.

African Soil: Obstacles to Be Overcome

In general the mother rocks of African soil are predominantly sandstone, metamorphic, and granite rocks, which have few nutrients, especially when they have been profoundly altered and are acidic. In addition to this constraint (and apart from the rates of organic matter to be maintained) is the low retention capacity of chemical fertilizers, along with blockages, deficiencies, and so on. In general, the same is quite often true of the alluvial deposits derived from the erosion of these materials.

Vast alluvial plains and large amounts of volcanic soil or soil formed from decomposed limestone possessing natural fertility are a rarity on the African continent. The relatively low level of agricultural potential is even more marked in West Africa, owing to the considerable expanse of ferruginous duricrust outcroppings or duricrust covered with fairly dense soil, which is very much at risk of irreversible erosion.

The decline in fertility in a great deal of the land exploited more intensively, with short or no fallow periods, raises questions regarding the capacity of the soil to produce more and in a sustainable manner. The extensive knowledge and expertise of farmers in soil conservation and fertility enhancement are often overlooked by those who are supposed to be assisting them (Courade and Devèze 2006). However, in a number of projects, this knowledge is being harnessed and enhanced.

The high cost of fertilizers and problems in making fertilizer use profitable are obstacles to their application, which remains very low in Sub-Saharan Africa. Initiatives involving the use of organic fertilizer (in particular through farmer-livestock producer organizations) are often inadequate because of the amount of work needed to prepare and transport compost or manure. More generally, investment in soil conservation has sometimes been perceived by Sub-Saharan African farmers as imposing "external" techniques that add to their workload.

The widespread adoption of soil conservation and improvement practices will remain limited as long as new land is available for clearance. Nonetheless, there is a growing consensus regarding diminishing yield,[1] which underscores the link between fertility and yield. This fertility is largely dependent on the content of soil organic matter, which must exceed the 0.7 percent threshold so as not to be a constraint. When this figure falls below 0.3 percent, it signals a problem of accelerated soil degradation. A consensus is also emerging with respect to the limits of conventional solutions (the cost of mineral inputs in light of lower cotton and food prices; the negative effects of tilling on fertility in African agroclimatic contexts).

Indeed, the debate regarding Africa's soil is not over. The complex interactions between soil degradation, poverty, and prices on the international market should be explored further in order to justify large-scale investment in restoring fertility, along with the accompanying measures necessary (Koning and Smaling 2005).

Biodiversity: A Key Component of Natural Capital

For biogeographical and paleoclimatic reasons, biodiversity in Sub-Saharan Africa is certainly "lower" than in other tropical regions (such as the tropical Americas and Asia), with the exception of the unusually high level of endemism found in Madagascar. This biodiversity is, nonetheless, an essential component of natural capital and is on the decline.[2]

What is known as "natural" biodiversity (as opposed to domestic biodiversity, resulting from manmade activities) is rapidly eroding, in line with the strong demographic growth rate and the contraction and fragmentation of natural spaces (Lawes 2007). Sub-Saharan Africa has roughly 10 biodiversity hotspots (Myers et al. 2000), almost all of which are located on the coastal region from Southern Africa to East Africa, including the group of islands to the southwest of the Indian Ocean. Added to these are the Congo-Nile crest landforms and the coastal Guinean forests. Two "wild regions" have been identified, namely, the Congolese basin and the Miombo-Mopané forest savanna complex (on the south banks of the Congo basin). To a large extent, the availability of traditional medicinal plants used by the vast majority of inhabitants in both rural and urban areas is determined by this "natural" African biodiversity.

Protected areas are one—but not the sole—priority area for preserving biodiversity (box 3.2).

BOX 3.2

Problems Associated with Protected Areas

With the growing international concern over the decline in biodiversity, protected areas were developed quickly and at an early stage in East and Southern Africa, where such land is currently thought to account for 14.5 percent of the surface area, and less so in West Africa, where this figure peaks at 8.7 percent. Despite the progress made in identifying hotspots, no real consensus has been reached about the process for prioritizing protected areas (Possingham and Wilson 2005, Orme et al. 2005). This process is driven more by opportunities as they arise rather than by well-founded scientific reasoning. New indexes (Forest et al. 2007) will, however, be able to contribute to better prioritization of conservation efforts.

The term *conservation* covers starkly different realities, depending on intentions and situations, given that protected areas are still subject to manmade pressures. At a number of major fauna reserves, the conservation refers to conscious management aimed at biomass optimization for the benefit of major herbivore mammals and their predators (Hayward et al. 2007)—sometimes to the detriment of diversity in the strict sense of the term—using practices suited to managing the environment and controlling fires. In the forest regions of the Congolese basin (and also in Madagascar), international nongovernmental organizations are becoming increasingly involved in the establishment and management of protected areas.

Regardless of the conservation models implemented and taking into account their current widespread nature, protected areas cannot serve as the sole tool for ensuring the maintenance of biodiversity in Africa. In fact, the maintenance of viable communities of flora and fauna in rural areas also determines the viability of protected spaces threatened by lack of access, fragmentation, and all manner of peripheral pressures (Clerici et al. 2007). This observation is linked to the issue of the integration of protected areas, which must be viewed jointly at the following three levels:

- Territorial integration has rarely been satisfactory, and few protected areas have really played a pivotal role in the organization of territories, except in three instances: when they serve as the main engine of the local economy, when they have led to joint initiatives and later impacted the organization of routes and transhumance activity, and when they have facilitated collaboration on transboundary problems that are often fraught with conflict ("peace parks").
- From a policy and institutional standpoint, the disconnect between conservation initiatives and poverty reduction efforts is long-standing. Discussions and attitudes

(continued)

BOX 3.2

(continued)

range from positions that are sometimes radical, or in any event divergent, and that can be broken down as follows (Adam et al. 2004): no link should be made between the conservation of endangered species and poverty; poverty reduction among the populations bordering protected areas leads to more efficient conservation; conservation should at least allow for coverage of all the opportunity costs associated with parks, inasmuch as conservation does not exacerbate poverty; and conservation strategies are based on the sustainable use of resources in the context of a poverty reduction strategy. In practice, everything seems to come down to priorities. Debates between strongly held positions based on theory or principle, sometimes viewed as irreconcilable, have quite often been settled merely by taking local situations into account and gradually implementing management procedures linking, at the local level, different levels of equilibrium between two imperatives. Progress is achieved with political and institutional integration through heightened awareness on the part of environmental institutions and the rapid strengthening of international commitments. At the state level, however, this integration has yet to be developed with respect to the linkage between deconcentrated services, decentralized authorities, and park administrations. Last, this integration has not yet been achieved in the area of intersectoral poverty reduction policies, which generally do not include conservation issues or local benefits derived by the populations, let alone ecosystemic services.

- Cultural acceptance of protected areas remains very uneven, despite the widespread nature of participatory approaches, which are often "ritualized" with the implicit "collusion" of populations or prominent figures. It is not unusual to see greater subjugation of the poorest to decisions made by village assemblies, where such persons traditionally have no voice. From a collective standpoint, it is also difficult to erase memories of plundering, which still color how populations perceive protected areas.

Furthermore, the sustainability of conservation tools is also linked to the political stability of states, and parks rarely survive armed conflicts or prolonged periods of political instability (for example, the Akagera-Mutara park in Rwanda). The development of conservation initiatives and transboundary management is likely to facilitate greater solidarity among states with respect to conservation measures, as a result of their regional dimension.

Funding the operation of protected areas and their opportunity costs is the subject of heated discussions, which, beyond the financial instruments themselves (trust funds similar to the predecessor of the Peruvian Trust Fund for National Parks and Protected Areas), are currently oriented toward payment for environmental services or establishing possible "conservation concessions." These solutions are the subject of spirited debate and accompanied by the real risks of the "hostage-taking" of natural ecosystems by populations and these populations' apathy toward management and development initiatives, or quite simply by the risk of an erosion of ethical values related to conservation and sustainable development, shaped in large measure by the emergence of "nature-based econometrics," which are sometimes simplistic or naive.

Domestic biodiversity (plant or animal) is also on the decline. By way of example, in the east-central part of Mali, 25 percent of sorghum strains have already disappeared from the Sahelian zone and production of 60 percent of the sorghum strains has already ceased in the Sudano-Guinean zones in the south. The same applies to local ruminants, whose age-old selection has led to adaptation to multiple climate, food, and health constraints specific to the African continent. The challenge that lies ahead is the preservation and improvement of this agrobiodiversity, in particular through partnerships between researchers and farmers within joint programs aimed at plant creation and selection, in light of population growth and climate change in the future.

Water: A Resource to Be Harnessed and Valued

Scarce in the Sahel and abundant in the humid forests, water mobilization and use vary widely from one location to another. Everywhere, water is an important resource to be harnessed to manage the sustainability of ecosystems and improve agricultural production. Integrated water management of resources such as the major Sahelian river basins should also facilitate optimization of the concurrent use of a scarce resource by using more appropriate practices to reduce the significant externalities currently generated by irrigated agriculture.

Multiple techniques are used in the case of plots. Mention should be made at this juncture of the efforts under way to limit erosion caused by rainwater through conservation agriculture, with a view to better mobilizing this agriculture through direct undersowing, in a bid to manage its use more effectively in the context of irrigated areas or drip irrigation.

Conservation and Development of Natural Capital

A long-standing debate has existed between persons who sometimes hold radical views of conservation and those who support the sustainable management of ecosystems (or *terroirs*) and natural resources. The management of forest ecosystems in particular is a controversial subject, depending on the degree of importance attached to technical and scientific considerations versus local sociopolitical constraints (box 3.3).

The issue of striking a balance between conservation and natural resource exploitation is applicable to all ecosystems. An additional factor is the strategy adopted by local actors generally confronted by global challenges that often exceed their capacities. African ecosystems, particularly soil, are fragile; hence the importance of avoiding irreversible forms of degradation (Giraud and Loyer 2006).

The development of this natural capital entails the optimization of the biomass production of soil, which can take a variety of forms. This is evidenced by the emergence, still in a fledgling form, of new facets of African landscapes (agri-urban zones, enclaves of land enclosed by cashew or various fruit trees,

BOX 3.3

Forest Extraction in Central Africa

The forest extraction landscape in Central Africa has evolved greatly over the past 10 years, initially as a result of forest policy reforms spearheaded mainly by the World Bank and later by the development of the rapidly expanding Asian markets (China and, before that, Japan) and the establishment of Asian enterprises in countries when regulations governing forest extraction are limited. Some progress has been made, however, such as the increase in the local processing of wood, higher taxes, and in some cases, higher forest revenue for states and, of course, the sustainable development of forests, which currently cover 60 percent of cultivated land.

The socioeconomic effects of forest extraction (Mengue Medou et al. 2005) are also being better managed. The rapid development of ecological "certifications" issued by third-party entities is in all likelihood a further favorable factor in mitigating the environmental and social externalities of forest activity and even perhaps in ensuring greater sustainability of the resource exploited. Certification is driven by the demands of consumers with high levels of awareness of the issue. More importantly, it is a recognized way for forestry companies to meet the commitments made with respect to the sustainable management of their concessions. It is impeded when consumer markets are not sensitive to quality and sustainability criteria (Southeast Asia is the leading importer of African wood).

The issue of the manifold impacts of forest activity on the composition, structure, organization, functioning, and sustainability of forest ecosystems remains, however, an open one. Specifically, it is often difficult to determine whether sustainable management initiatives are applicable to wood as a resource or to the entire ecosystem. Beyond acknowledging the sometimes severe environmental damage (human penetration, opening of roads, selective cutting and modification of undergrowth, hunting, poaching, and the like), is it sufficient to limit the disturbances caused by forest activity, along with the conditions for regeneration of the forest system, in order to address the issues hammered out in international circles?

---

private forest plantations, community forests),[3] which are relatively new initiatives, especially since they are often homegrown and have received little external support. Such initiatives point to the need to examine these varyingly scoped organizational models, ranging from the local models driven by incipient decentralization efforts to the transnational models that necessarily include management of major river basins, the organization of transhumance routes, and the like. This support and territorial reciprocity should also materialize by addressing the actual integration and territorial inclusion of protected areas in regional spaces. This requires establishing systems and networks that incorporate the various levels of biodiversity conservation, such as regulated protected areas, traditionally protected areas, and natural microsystems found in rural spaces (such as gallery forests and fallow, forest patches at the foothills of duricrust and rock outcroppings). The specialized used of space (fencing) based on the reintroduction of adapted plant materials, integrated into an organized land system, can also contribute to this effort by fostering the (re)diversification of rural landscapes.

In the Sudano-Guinean transition zones, tall grassy savannas form an extremely important layer of biomass that is currently underdeveloped (standing fodder reserve partially used by animals). Because this biomass is not adequately exploited, it feeds major brush fires. Given the setbacks encountered in storing natural hay obtained from grasses that grow very quickly at the start of the rainy season, fodder crops are being introduced—with some difficulty—in regions where food is grown mainly for subsistence purposes.

In arid zones, in view of developing sparse, random, and dispersed biomass, nomadic pastoralism is irreplaceable: the techniques used by nomadic pastoralists are now recognized as being very sophisticated, and no alternatives have been proposed to develop this pastureland. Here again, human capital revolves largely around the continuation of these adapted traditional activities. The end of regional conflict and the establishment of interstate management and negotiation instruments, driven by the appropriate pastoral water improvements, should not only facilitate the spatial organization of these mobility-based activities but should also contribute to minimizing conflicts between farmers and herders and thus directly contribute to the conservation of protected spaces, which form the last bastions of nomadic pastoralism in areas that are heavily and increasingly used for agricultural purposes. The issue is therefore one of territorial development, the design of which should be consistent with procedures under the New Partnership for Africa's Development.

Three Cross-Cutting Issues to Be Considered

Climate Change: Managing Uncertainty

Climate change has made it possible to place greater emphasis on the long-term outlook in development-related discussions. This changed outlook has the merit of requiring greater consistency and convergence among the different approaches, including adapting to climate change, food security and poverty reduction, maintaining biological and cultural diversity, and combating desertification. The initiatives that cover these different issues link the preservation of local public goods with global public goods.

Currently, there is recognition of the fact that climate change will have a major impact on Africa, probably through the increased frequency and intensity of unusual events (droughts, heavy rains, and cyclones in the regions affected). In light of this, the people of Africa are highly vulnerable given the lack of supplies, low levels of land irrigation, and limited research and development capacities. It bears noting, however, that climate events have always been part of the daily lives of African farmers and that, in this sense, adaptation to climate change is not new. However, collective action, which to some extent nurtures an adaptive spirit (Giles 2007), is being undermined by the shift toward the individual privatization of farming.

It must be acknowledged that adapting to climate change also offers an opportunity to place long-standing issues on the agenda once again (such as combating desertification, soil protection, and controlling fires). This process also offers an opportunity to step up the work already done in the areas of water and soil conservation, sustainable irrigated agriculture, or even the establishment of insurance-related mechanisms (agricultural, food, seed insurance), given the now pressing nature of these issues. However, adaptation will require the strengthening of the climate monitoring system at the level of the entire continent, with the opportunity for significant improvement in the area of greater synchronization of farming practices with meteorological events (optimization of seeding dates). Consideration of the impact of climate change on biodiversity is another strategic issue that will become an important dimension of conservation policies, despite the fact that, here again, current knowledge is recognized as being highly inadequate (Thuiller 2007).

Energy: An Optimal Service to Be Provided

Of the energy services linked to agriculture, mention will be made only of the following, which are considered important in Sub-Saharan Africa:

- Mobilization of the appropriate sources of energy needed for agricultural production and its development. Agriculture involves labor and transportation (such as manure, organic matter, harvests, inputs). Much remains to be done in the area of improving use of animal energy.
- Production of energy that can be used by farmers or other consumers. The production of agrofuels may prove profitable and may offer an energy option that could be satisfactory in tropical regions, as was the case in Brazil, when it does not entail the systematic conversion of natural spaces into agricultural land or create competition between these potential agricultural spaces and food crops. The development of marginal land on an experimental basis for this purpose is planned in several countries in the Sudanese zone.
- Provision of the energy needed for cooking. Wood remains the main source of energy in Sub-Saharan Africa.

Overall, agriculture needs to be repositioned in an energy system that should provide optimal service from the standpoint of resources, economic and social costs, and protection of the local and global environment (Laponche 2008).

Agricultural Production: Yields

Greater agricultural production in broad terms must take place in the context of a change in the rational use of land, based on the establishment of new balances between cultivated land, woodlands, pastureland, and protected land. New models of sustainable development must be adopted to address environmental constraints, demographic growth, or population needs, taking into account ecosystem management methods that, thus far, have been able to preserve natural capital.

The strong tendency to increase cultivated land areas on the basis of short-term strategies designed to meet immediate needs leads to limits that have already been reached in the case of some Sub-Saharan land. However, there are a number of interesting solutions for stabilizing cultivated land while increasing production. Consequently, this shift in agriculture toward greater production and diversification must be managed from a spatial standpoint.

BOX 3.4

The Doubly Green Revolution and Undersowing: Standing at a Crossroads

The introduction of undersowing techniques offers numerous advantages for the "ecological intensification" of production systems. Brazil is a noteworthy example of a country where these systems have been developed with positive effects in the environmental sphere (such as increased organic matter, stimulation of biological activity, reduced runoff, and greater water penetration into the soil) and in the economic or social spheres (fewer inputs, safe yields, less taxing work—particularly weeding, which is often done by women in Africa—and the provision of fodder). Although these undersowing systems are being adapted to Africa, including Madagascar, several constraints need to be addressed:

- The first is the selection or even introduction of plant cover materials suited to the long dry season and recurring rainfall events in the Sudanese zone, as well as access to seeds, given that many of the best adapted species are still expensive.
- The second pertains to the understanding, for each parcel, of technical pathways that are still complex and the integration of these innovations into the farming system (use of fertilizers and herbicides, crop rotation, means of transport and tools, and the like).
- The third, which is problematic in the context of traditional livestock activity (particularly in regions that host transhumance activity), pertains to controlling roaming livestock. This requires enclosing or fencing parcels of land used for undersowing and thus the establishment of provisions for the special use of space to control brush fires.

(continued)

BOX 3.4

(continued)

Other issues relate to correcting soil deficiencies, producing seeds from cover plants at costs compatible with the financial yield of the land, and the use of environmentally hazardous weed killers.

While seemingly simple, the dissemination of these innovations, in fact, requires a significant effort to train farmers. However, its strength lies in the fact that it does not call for heavy equipment, such as tractors, or the precipitous and widespread use of fertilizers that the poorest cannot afford, or complex agrarian reform (with the exception of a number of land problems to be resolved in a traditional context). However, this process calls for the rethinking of customs (common grazing land). It is also an appeal for the development of space (watershed management) with access that must be made secure (land rights, leases, hedges). The challenge is therefore one of organizing the social change necessary for the large-scale dissemination of these innovations, which serve as an example of sustainable development—in the economic sphere, by increasing yield; in the environmental sphere, by halting erosion and improving soil fertility; and in the social sphere, by investing in human capital, namely, farmers.

An important discussion revolves around the type of agriculture that facilitates both an increase in production and the preservation of ecological potential. Given the limits of mechanized and chemical-dependent agriculture, Michel Griffon (2006) promoted the concept of the doubly green revolution (see box 3.4), based on production technologies that draw on the functioning of natural systems and on agricultural policies that facilitate their dissemination among farmers. The doubly green revolution encompasses the notions of agroecology, eco-agriculture, and conservation, while incorporating the concepts of economic viability and social equity.

Generally speaking, the challenge therefore lies in the ability to embrace all opportunities for increasing biomass production and use in more diversified agricultural systems that are not too input intensive, where the insurance-related value of biodiversity has been preserved. Taking local situations into account should help settle the debate between supporters of "green" and "doubly green" revolutions.

Adapting Public Policies to the Ecological and Human Situations of Territories

The significant production potential of Africa's biomass cannot conceal its vulnerability: resources to be shared in increasingly contentious situations, careful management of the exploitation of woody flora and the replenishment of forest reserves, soil erosion and the fragility of living landscapes in the wake of deforestation, and management of water resources in the context of climate change. Options related to the capacity of ecosystems to reconcile food or energy production with the conservation or reproduction of ecological capital should first be discussed with local stakeholders and settled at the policy level.

The management of ecosystems over time, taking into account environmental and higher production imperatives, constitutes a cross-cutting dimension of public policy. African countries tend not to accord priority to this topic given their myriad other immediate concerns. For this reason, first-generation Poverty Reduction Strategy Papers virtually overlooked the need to preserve natural capital, opting instead for sectoral approaches that make it easier to quantify performance. While progress has been made in the area of environmental economics, the understanding of environmental goods, costs, and services should not be confined solely to what is easily quantifiable. Indeed, a commercial value cannot be assigned to all these goods and services. Their evaluation methods have not yet been firmly established, and agreement to pay the relevant costs remains hypothetical. The problem of supporting environmental public policies in a manner commensurate with the challenges, followed by their implementation, remains unresolved in Sub-Saharan Africa.

For the time being, emphasis has been placed, largely by the international community, on the protection of nature and the preservation of forest areas. Recent Climate Convention discussions related to the consideration of land clearance, which has been sidestepped in carbon assessments, demonstrate the progress made to support international commitments through incentives adopted in the context of the REDD initiative. Much remains to be done in this area, as well as in the areas of better water and land management and the implementation of energy policies.

Greater inclusion of the territorialization dimension in public policies, taking into account how farming societies use nature, is required to adapt public actions to the specificities of territories and natural spaces. The recognition of legal pluralism (see chapter 13), in particular of local or custom-based provisions that are accepted and in line with sustainable management objectives, offer a sound basis for the work of local stakeholders involved with voluntary sustainable management initiatives.

Last, consultations aimed at drafting and implementing public policies should facilitate the determination, with input from the relevant stakeholders, of why and for whom ecosystem management is important in the long term. Such an approach should avoid the "tyranny of small decisions" that has too often typified rural development in Africa, by imparting meaning and consistency to long-term activities.

Notes

1. See the example of cotton yield, which, after peaking at close to 1.4 tons per hectare of seed cotton, is now approximately 1 to 1.1 tons per hectare.
2. See *Afrique contemporaine,* issue 222, on natural resources.
3. Often the result of the recognition of land rights, which is accompanied by tree planting (preferably of fruit trees or exotic species).

Bibliography

Adam, W. M., et al. 2004. "Biodiversity Conservation and the Eradication of Poverty." *Science* 306: 1146–49.

Balmford, A., et al. 2001 "Conflicts across Africa." *Science* 291: 2616–19.

Clerici, N., et al. 2007. "Increased Isolation of Two Biosphere Reserves and Surrounding Protected Areas (WAP Ecological Complex, West Africa)." *Journal for Nature Conservation* 15: 26–40.

Courade, G., and J.-C. Devèze. 2006. "Des agricultures africaines face à de difficiles transitions." *Afrique contemporaine* 217.

Forest, F.. et al. 2007. "Preserving the Evolutionary Potential of Floras in Biodiversity Hotspots." *Nature* 745: 757–60.

Frost, G. H., and I. Bond. 2008. "The CAMPFIRE Program in Zimbabwe: Payments for Wildlife Services." *Ecological Economics* 65 (4): 776–87.

Giles, J. 2007. "How to Survive a Warming World?" *Nature* 446: 716–17.

Giraud, P., and D. Loyer. 2006. "Capital naturel et développement durable en Afrique." In *À quoi sert d'aider le Sud.* ed. S. Michaïlof. Economica, Paris.

Griffon, M. 2006. *Nourrir la planète.* Odile Jacob, Paris.

Hayward, M. W., et al. 2007. "Carrying Capacity of Large African Predators: Predictions and Tests." *Biological Conservation* 139: 219–29.

Intergovernmental Panel on Climate Change. 2007. The Fourth Assessment Report. Geneva, Switzerland.

Jullien, F. 2006. "Nomadisme et transhumance, chronique d'une mort annoncée ou voie d'un développement porteur?" *Afrique contemporaine* 217.

Koning, N., and E. Smaling. 2005. "Environmental Crisis of 'Lie of the Land?' The Debate on Soil Degradation in Africa." *Land Use Policy* 22: 3–11.

Lawes, M. J. 2007. "The Effect of the Spatial Scale of Recruitment on Tree Diversity in Afromontane Forest Fragments." *Biological Conservation* 139: 447–56.

Laponche, B. 2008. "Prospective et enjeux énergétiques mondiaux et nouveau paradigme." Working Document 59. AFD, Paris.

Mengue Medou, C. et al. 2005. "Evaluation des impacts socio-économiques: cas d'unité forestière d'aménagement de la compagnie forestière Leroy-Gabon." *Vertigo* 6 (2): 1–8.

Myers, N. et al. 2000. "Biodiversity Hotspots for Conservation Priorities." *Nature* 403: 853–58.

Orme, C. L., et al. 2005. "Global Hotspots and Species Richness Are Not Congruent with Endemism or Threat." *Nature* 436: 1016–19.

Possingham, H. P., and K. A. Wilson. 2005. "Turning Up the Heat on Hotspots." *Nature* 436: 919–20.

Sankaran, M., et al. 2005. "Determinants of Wood Cover in African Savannas." *Nature* 438: 846–49.

Thuiller, W. 2007. "Climate Change and the Ecologist." *Nature* 448: 551–52.

World Bank. 2006. "Where Is the Wealth of Nations? Measuring Capital for the 21st Century." Washington, DC.

———. 2007. *World Development Report 2008: Agriculture for Development.* Washington, DC.

Chapter 4

Entangled Issues in Need of Clarification

Jean-Claude Devèze

The manner of approaching the problems of, on the one hand, limiting population growth and distributing the populace across towns and countryside taking into account agricultural potential and, on the other, of increasing productivity to address the demand for food while sustainably managing ecosystems will be decisive for the future of African agriculture. What virtuous circles might be promoted among these key factors that would to steer clear of the vicious circles leading to underemployment, famine, population exodus, and desertification? Not only do the demographic, economic, and environmental dimensions of agricultural development have to be integrated, but also its social, political, and cultural dimensions.

The discussion of demographic, economic, and environmental challenges in the three preceding chapters has already shown the existence of links between these three areas in countries the majority of which have sizable agricultural populations, and where economic growth and employment remain highly dependent on an agricultural sector that is experiencing difficulties and on fragile ecosystems. The purpose of this chapter is to recall the importance of the political, social, and cultural issues at stake and then to clarify how all these stakes come into play on different scales, particularly on the scale of the farm itself and that of the subcontinent. Finally, an attempt will be made to place these various interdependent issues in perspective and to quantify them, to extract from this entanglement the major problems of Sub-Saharan agriculture.

The Various Types of Issues at Stake

The first three chapters showed the importance of the stakes at play on the demographic, economic, and environmental levels. These involve regulating the demographic dynamics of a young population that is unevenly distributed in space, promoting economic dynamics that are built upon improving the productivity and diversification of agriculture to ensure that it is a good fit with buoyant markets, and finally taking due account of the environmental imperatives of sustainably managing natural resources and ecosystems.

The political dimension also proved to be crucial during the study of these issues. Thus, although the agricultural world is still very much removed from power at the national level, a first political challenge is to strengthen the checks and balances in rural areas (Devèze 1996). Like many other social categories, farmers can become more autonomous and achieve greater power by promoting their organizations and easing their demands. The problem of the political weight of farmers is partially associated with the promotion of democratic life at all levels, including in the rural communes and communities. Views are still cautious about the possibility open to small farmers to pursue counter-hegemonic projects (Bayart, M'Bembe, and Toulabor 1992) and on the role that democratic ascendency might have (Étienne 2007) in development at the current stage. Will the democratization and decentralization of power and the structuring and enhanced accountability of professions grow together? It is difficult to sketch out the possible future of democracy in Africa (Mbembé 2000) in view of the disruptions of political life that is all too often situated somewhere between coups d'état and elections that are postponed or rigged, between social movements and fights for identity.

Not only are African states at pains to put a democratic life in place and to define policies but also most of them lack, or use poorly, the financial resources that are essential for developing their economies and improving their territories. It bears recalling that Sub-Saharan Africa has been and remains partially dependent on external support, ranging from food aid to budget financing, thus giving rise to a second political challenge, that of emerging from dependency and its multiple unfortunate consequences.

External financing, which for agriculture comes primarily from official development assistance (ODA), has not been sufficient in many countries and small regions to favor dynamics that substantially benefit their populations and make it possible to put in place the infrastructures and institutions that will foster development, especially in the countryside. Concerning Sub-Saharan agriculture, Gilbert Étienne speaks of flagrant failures alongside a few successes, such as cotton in West Africa, hybrid corn in East Africa, or cut flowers in Kenya. In Sub-Saharan Africa the green revolution has been quite marginal by comparison with Asia for various reasons, such as the limited number of ideal rainfed plains in Africa, weaknesses in cooperation approaches, inadequacies on the part of public institutions and policies, and poorly conducted free-market approaches (Michailof 2006). A fourth political issue at stake is therefore the place and role of ODA in the complex set of interactions between countries of the North and of the South (Charnoz and Severino 2007). To improve the performance of ODA, is it sufficient for the donors to have undertaken—first in Rome (2003) and then in Paris (2005)—to bring relations with the developing countries back into balance in terms of ownership, alignment, and the harmonization of donor aid?

The overlapping of major policy issues for the future of our planet raises the question of the integrated management of global affairs in the context of power relationships based on local dynamics (Hourcade 2008) and on the economic and political weight of the various nations and regions.

While the developed countries have spent several centuries laying the groundwork for their development and two centuries managing their demographic transition, the countries of Sub-Saharan Africa must meet challenges of an unprecedented scope in an international context that is particularly difficult for them:

- Industrial development in the older and especially in the newly industrialized countries makes it more difficult for new countries to enter the game.
- Delays in education and research are limiting capacities for innovation, and underinvestment in rural infrastructure is putting a damper on progress.
- Pressures in favor of opening up markets and raising agricultural prices are impediments to introducing temporary protections, even when this has historically been a precondition for take-off, in Europe and the United States, in Japan, and in the newly industrialized countries, thus leading to problems in strengthening local economic fabrics.
- The energy crisis calls back into question the past strategies based on ever-increasing consumption of low-cost fossil fuels, making it necessary to invent novel alternatives that are consistent with the standards of living of the population and with environmental imperatives.

To overcome their initial disadvantages, the countries of Sub-Saharan Africa must mobilize all their forces around a project to achieve credible development that takes the potential of their agricultural systems into account. This assumes that economic choices are not dictated by the various sources of financing, that social regulation systems are not co-opted by the interplay of interest groups associated with the "elites" (Meisel and Ould Aoudia 2008), and that political leaders win the trust of farmers while steering clear of demagogic promises (such as allocation of land tenure by the government as part of some major offensive or doubling the production of rice).

Regaining policy ownership in this manner is all the more important in that the poverty reduction effort in rural Africa is at a standstill and inequalities are on the rise, both among farmers and between cities and rural areas. Furthermore, the increases in food prices are likely to accentuate even further the problems of poor farm families that need to purchase foodstuffs to bridge the gap before their harvests come in. Finally, access to land is unequal in many regions (see chapter 6). One major social issue at stake is therefore to decrease poverty and inequality in rural areas, which should contribute not only to preserving social cohesion but also to accelerating the demographic transition and diminishing the abusive exploitation of natural resources.

Yet another socioeconomic issue is that of agricultural employment, which takes a number of different forms. For family farms, the objective is to make the best use of their available labor, to address the problems of peak-work periods and periods of underemployment, and to deal with the organization and compensation of labor and ultimately the status of women and youth. For agribusinesses, the issues also involve improving the mobilization of the necessary labor force, but, additionally, there are specific problems pertaining to the qualifications, wages, living conditions, and status of agricultural workers. The future role of agroindustries and their hiring possibilities are highly variable, depending on the production basin concerned (box 4.1). There are also delicate issues relating to the significant recourse to child labor and to developments with respect to the work of women, who more often than not prefer to cultivate their own parcel rather than work for the head of household or their husband.

A major sociocultural issue in play is that of the climate of trust or mistrust prevailing in rural areas; it is decisive with respect to decision making, for example, when questions of innovating, investing, or migrating arise. This involves the all-too-frequent scorn shown toward small farmers by the leadership classes, and the poor example given to youth by those who enrich themselves too easily and fail to promote a long-term commitment to living off the land through hard work. Of course, all this is interconnected with the training, apprenticeship, and information mechanisms for farmers, which have been, by and large, overlooked (see chapter 10 on human capital). An even more delicate matter is that of analyzing the consequences of traditional cultural practices, in particular, those governing relations with others, which are often characterized by respect for elders, a search for consensus, and the manner of managing risks and fears, such as the fear of being poisoned by one's neighbor.

Confidence in the future may be linked to multiple factors, such as an education that frees up one's potential, or organizations that make work activities more secure, or leaders who shed light on the choices that are possible. Might not the emergence of a territorialized project (Deffontaines and Prod'homme 2001) also be capable of sustainably mobilizing local stakeholders affected by the accumulated mass of inputs and influences within a sensitive context of confrontation between endogenous dynamics and external interventions, between urban and rural spaces, between local influences and increasing confrontations with an ever more present modernity (radio, television, mobile telephony, Internet)?

BOX 4.1

Typology of the Agricultural Systems and the Place of Agribusinesses in the Senegal River Delta

As an approach to the difficult issue of the diversity of agricultural systems and the place of agribusiness, the case of the Senegal River Valley delta can be instructive. In the delta, there are family farms on 40,000 hectares, agroenterprises on 10,000 hectares (some being essentially family run), and three principal agroindustries: the Société des conserves agricoles du SENEGAL (SOCAS), the Compagnie sucrière du Sénégal (CSS) and the Grands domaines du Sénégal (GDS). A typology of the different agricultural systems present in the delta can be made on the basis of three main criteria: the share of production that is marketed, the share of family labor, and the amount of equipment used.

Family agriculture covers both family farms that are more or less capable of evolving depending on their initial situation, and a portion of the agroentrepreneurs, namely, those with a mainly family-based labor force. The remaining agroentrepreneurs function more like a conventional business, with a head of operations and wage earners. However, the distinction between family farms and agroentrepreneurs is not clear-cut, because some long-standing traditional farms have now become genuine agricultural enterprises.

Among the agroindustries, only SOCAS maintains strong contractual ties with producers, within the framework of an interprofessional tomato association. It buys tomatoes from organized producers, with purchases amounting to 69,000 metric tons in 2005–06 and 50,000 metric tons in 2006–07. It then processes them into tomato concentrate for the Senegalese or other nearby markets. But SOCAS is facing increasing competition from a Chinese enterprise that reprocesses imported concentrate. The CSS (10,000 hectares) employs a permanent labor force (3,000 persons) as well as temporary labor (2,000 persons), some of whom come from southern Senegal. The sugar produced (90,000 metric tons) is sold primarily in the region. The GDS exports greenhouse-grown cherry tomatoes (47 hectares) to Europe, using *Compagnie Fruitière* boats; it also seeks to produce tomatoes and green beans in open fields.

Source: For SOCAS: *www.inter-reseaux.org/IMG/pdf/Senegal_Note_Analyse_Tomate_n1_CGERV-2.pdf.*

The most overarching problem posed is that of the manner in which societies decide upon their disappearance or survival or their development on the basis of how they engage in relations with their neighbors, settle environmental problems, and mobilize around leaders to meet the challenges that arise (Diamond 2006). Has tropical Africa heard the call to become an agricultural giant (Gourou 1991)?

Issues to Be Analyzed on Different Scales

Another way of looking at the stakes associated with the future of African agriculture is to review them on different scales: local family farms, agrarian systems, the domestic agricultural sector, the broader region, and the subcontinent. This differentiation should make it possible to make a better assessment not just of the demographic, environmental, and economic stakes but also of the social, political, and cultural stakes. In the following, we examine only the first and last levels: the individual farm and the subcontinent.

For local family farms, the major challenges for the future are of various orders (Devèze 2004; Brossier 2007).

Most of these farms, which are sparing in their purchases of inputs and energy, have a substantial capacity to resist external random events so long as natural resources remain available and ecological equilibriums are preserved. In contrast, given the rapid changes now under way on our planet, these farms could rapidly find themselves marginalized, owing to their lack of competitiveness vis-à-vis more productive modern agricultural systems that are better subsidized and, above all, benefit from a more secure economic environment. The issue at stake is, therefore, economic. It comes down to transitioning from family units with fragile activities toward agricultural operations managed as small businesses capable of producing more and better in order to sell agricultural produce sought by the market; farms react extremely well to prices and are capable of increasing production when prices rise (Gafsi et al. 2007).

The issue becomes ecological when farmers are obligated to overexploit natural resources, the soil in particular (Giraud and Loyer 2007), to ensure the survival of their families, or when the introduction of selected varieties calls into question the possibilities afforded by biodiversity.

Small farmers, as distinct from agricultural entrepreneurs who look upon themselves as business heads like any others, live within family activity units. Income streams are not broken down into categories such as wages, profits, and rents. Limiting risks is of primordial importance to survive and, if possible, to preserve some degree of autonomy. The primary value continues to be labor. What is at stake is thus cultural: it is a question of finding the proper niche between tradition and modernity, between liberating individual capacities and engaging in family and community solidarity.

The issue is also social, because departing from local conformism by marking one's differences poses the risk of being cut off from neighbors; moreover, the more rapid enrichment of those who are more dynamic or more influential is often the source of inequalities.

The issue may be demographic when the marriage possibilities of the young depend on the family, when the value added by females is associated with the children they bear, and when the question of leaving or staying on the farm is influenced by job and earning opportunities.

Economic and social changes on farms also depend on the strength of relationships and on struggles for power that relate to land and the complex interplays of various stakeholders. Hence, in the final analysis, the issue is a political one.

On quite a different scale, it is important to examine how agriculture in Sub-Saharan Africa is a stakeholder in major issues at the global level.

Taking account of the demographic characteristics of rural Sub-Saharan Africa, the three following problems are raised:

- How should population policies be conducted when African states lack the authority and resources to implement them (see annex 1B)?
- How can meeting the essential needs of a growing rural population best be addressed?
- How should we approach the training of the many young children of farmers until they become autonomous, and what kinds of jobs should be proposed to them?

On the economic front, it is difficult for African and Malagasy agricultural subsectors, except in the case of certain products that benefit from a favorable ecological situation (such as gum arabic and vanilla), to consistently rank among the most competitive. Priority should therefore be accorded to strengthening their competitiveness and improving their organization and economic environment.

With regard to food supply, Africa is the continent that must face the greatest challenge, as the multiplier for satisfying its increasing food needs between 2000 and 2050 is on the order of 5, as compared to 0.91 for Europe (Collomb 1999). This factor of 5 is explained by anticipated population growth but also by the need to improve the current caloric ration and by the anticipated diversification toward meat-based diets. This would be expected to contribute to increasing pressure to provide foodstuffs, as reflected in the rise of prices on the world market, which is further compounded by the rise in transport costs. Sub-Saharan Africa should meet this challenge (Blein et al. 2008) by being capable of responding to growing regional demand for foodstuffs at prices that provide better compensation for the efforts put forward.

Meeting the challenge of feeding growing poor populations becomes even trickier in a period of increasing volatility in commodity prices. Taking as an example the market for the three main grains (corn, rice, and wheat), we note a sharp rise in prices since 2007, which has even led to food riots in some large cities. The many different consequences of pressures in the foodstuff markets need to be examined in greater detail, in particular the largely overlooked consequence of how increases in world prices have impacted the prices paid to the African small farmer. Unfortunately, it must be observed that prices are rising more rapidly and sharply for the urban consumer than for the African producer, because the authorities have so far been more concerned with finding short-term measures to ease the burden of urban dwellers than with the implementation of policies to support the productivity efforts of farmers by ensuring their equitable compensation. Finally, we must not overlook the fact that poor small farmers purchase a portion of their food and thus lose rather than gain more from the increase in food prices.

On the social level, agricultural underemployment prevails in many farms, in particular in the dry season, and temporary migration and nonagricultural activity still fall short of offsetting this problem. Moreover, it is widely recognized that women and young people often perform the least desirable work, such as land clearing, resetting rice, or the manual harvesting of cotton. Finally, there is increasing social differentiation, as illustrated in studies on the typologies of farms and their evolutionary paths (Gafsi et al. 2007).

With regard to energy, the agricultural sector of Sub-Saharan Africa is less affected by rising energy costs than are those of the developed countries, owing to Africa's low consumption at the production level (extremely low level of mechanization, very few petroleum-based inputs, and little refrigerated storage). Sub-Saharan Africa does appear, however, less apt to find alternatives because of the weakness of its economic fabric and the scant research and development work carried out. The advantages of agrofuel production through short energy supply pathways making use of the sizable potential for biomass production in the tropics remain one avenue to be explored with a view to providing energy near at hand for certain agricultural activities.

In terms of greenhouse gas emissions, Africa is not at the top of the list as are the developed countries, given its relatively low level of carbon dioxide (CO_2) emissions. In contrast, the relations between rural dwellers and their natural environment raise multiple issues. For example, brush fires and the use of fire for clearing land are widespread, as shown by the aerial photographs of the planet in Al Gore's film *An Inconvenient Truth*. Moreover, the use of wood for cooking is a source of excessive pressure on trees in many regions, even though there are measures such as improved stoves to limit the use of fuelwood (Louvel and de Gromard 2007. In addition, cattle raising results in considerable methane emissions, but less so when livestock is pastured. Overall, agriculture emits more methane and nitrous oxide than it does CO_2 (Jancovici 2002); hence the importance of further studying the various consequences of the different greenhouse gas emissions in conjunction with the decrease or increase in biomass production, with deforestation and land-clearing or plantations, and with modes of livestock rearing and crop growing and the way these interrelate.

More important, the pressure on natural resources from human and animal sources increases in those regions of Africa that are most densely populated, and those pressures can then pose a threat to biodiversity and soil fertility. French agronomists (Piéri 1989; Angé 1991; Raymond and Beauval 1995; Dufumier and

Bainville 2006; Gigou et al. 2006; Jouve 2006; Billaz 2007; Bied-Charreton 2008; Calame 2008) have, each in his or her own way, expressed considerable concern about the trend in the fertility of many poor and fragile African soils, which are increasingly subjected to continuous cultivation using plowing techniques likely to favor erosion, without sufficient compensation for mineral exports (particularly of phosphorus and calcium) and without sufficient organic replenishment. Because agronomy is a science of localities, these problems need to be addressed at the plot level and on the ground (Kleene, Sanogo, and Viestra 1989) in light of multiple factors, such as the cost of fertilizer, capacities for the production and shipping of manure, the desire to limit the use of herbicides and pesticides, recourse to nitrogen-fixing leguminous plants, crop rotation, and the agro-sylvo-pastoral equilibriums to be preserved or promoted.

Finally, Sub-Saharan Africa, which is highly dependent on rainfed crops given the small proportion of land under irrigation (3 percent of the land under cultivation), is vulnerable not only to the vagaries of weather but also to climate change (Desanker 2007; chapter 3).

Major Problems to Be Set in Perspective and Quantified

Because of the poor quality of agricultural statistics in Sub-Saharan Africa and the extreme diversity of the ecosystems to be taken into account, studies focused on forecasting are difficult to carry out. In 2000 the Food and Agriculture Organization estimated the land areas under cultivation in Sub-Saharan Africa at 228 million hectares, representing 22 percent of cultivable surfaces. However, 20 percent of these cultivable surfaces are in need of protection to ensure that they are not degraded. Sub-Saharan Africa, like Latin America, has far more reserves of cultivable land than do other regions of the world but a smaller proportion of irrigable land.

According to the forward-looking Agrimonde study on "agriculture and food 2050" (works in progress being conducted by the French agricultural research organizations INRA and CIRAD, estimated yields in wheat-equivalent terms to allow for comparisons of calories produced by hectare are estimated at 1 metric ton in Sub-Saharan Africa, which is twice less per hectare than for the world as a whole. Between 1961 and 2003, wheat-equivalent yields per hectare doubled in Sub-Saharan Africa while, surprisingly, the areas under cultivation per agricultural worker dropped by nearly half; in Sub-Saharan Africa 10,000 kilocalories were produced per crop production worker, compared with 40 times as much in the most developed countries. Sub-Saharan Africa is thus intensifying its plant production, owing to the increased productivity of labor per farmer and in production per hectare, but at a slower pace than on other continents and from an extremely low starting point.

There is substantial potential for improving production in Sub-Saharan Africa, but the various characteristics of this potential need to be better understood for it to be more effectively developed, at the same time taking account of new balances that need to be found. Among these balances, particular mention must be made of the need to reach the optimal combinations between crop farming, livestock rearing, and forestry, between mechanization and motorization, between the use of chemical and natural fertilizers (Dufumier 2005), between carbon emission and capture, between plowing and direct seeding, between the production and consumption of energy, between irrigation and rainfed crops, between recourse to selected seeds for standardized food production and the use of biodiversity resources for varied food production (Calame 2008), between biological approaches and the use of phytosanitary products, and between specialization and diversification. New farming systems and agrarian systems that have been reconsidered in this light need to be refined by mastering innovations and supporting exchange networks.

Various hypotheses can be developed about how to resolve food, environmental, and energy problems at one and the same time (Griffon 2007). Trend scenarios, based on the continued expansion of the areas under cultivation and small improvements in yields, suggest that Sub-Saharan Africa has the potential to meet the growth in food demand from a quantitative but not qualitative standpoint (shortage of animal protein), especially while coming up against ecological impasses (deforestation, overexploitation of natural resources). The pronounced disparities in situations between agroecological zones (see chapter 3) should lead to migrations toward favorable zones that are sparsely populated. This, however, is a far from simple matter and may result in civil disorder. Against this background, Michel Griffon proposes a scenario for 2050, which is considered to be risky. It is based on a first objective of food production and a subsidiary objective in the longer term of biomass production, at the same time protecting a portion of the major forest massifs. In 2050, for Sub-Saharan Africa, this scenario could translate into food equilibrium and a positive balance in biomass energy of 240 million metric tons of oil equivalent, thanks to a threefold increase of the areas under cultivation or used for pastureland and a 25 percent rise in yields.

A study on West Africa (Blein et al. 2008) also proposes several different scenarios: the optimistic one, based on a favorable regional and international context, indicates only a 50 percent increase in land area under cultivation by 2030, but a twofold increase in plant yields and a 2.5 increase in animal production. In this scenario, the opportunity afforded by the increase in agricultural prices is fully taken up by half of the farms, the food supply improves, and the pressure on natural resources remains reasonable. These scenarios prepared for West Africa also address the following important issues:

- Irrigated agriculture and the development of the lowland basins will likely preclude a reduction of African imports of rice, given the cost of the improvements and their maintenance as well as the limited availability of water; hence, the importance of rainfed rice.
- The development of short-cycle livestock farming, milk production, and fish farming should make it possible to improve the food ration in terms of animal protein.
- In an international context favorable to trade from the developing countries toward the developed countries, increased exports of African agricultural products could limit the foodstuff production intended for local and regional markets.

To increase agricultural production by a factor of four or five in two generations, given population growth and the need to improve diets, it is necessary to have a better grasp of changes in the increase in yields (from 25 percent to 100 percent in the above hypotheses) and the land areas used (from 50 percent to 300 percent in the same hypotheses). Some experts tend to lay emphasis on increasing the land used, where this is still possible, given the difficulties of increasing yields (problems of soil fertility, the high cost of fertilizer, and the limits on possible transfers of organic matter). Others prefer to wager on increasing yields while preserving ecological equilibriums. In the past, African small farmers have tended above all to increase the areas they plant and boost herd size; future trade-offs made by farmers between the various possible strategies for making the most of their natural capital will have significant consequences for jobs, incomes, diets, and the environment.

In these various studies, the consequences in terms of agricultural employment and the marginalization of many agricultural households (50 percent? 75 percent?) that are unable to improve their performance are not dealt with in depth. This raises the thorny question of the social cost of agricultural transition, or, in other terms, the painful transformations of the marginalized family farms. Taking a social policy approach to dealing with this problem is difficult in countries that lack policies in these areas as well as the financial resources needed to implement them. As a result, it is necessary to envisage supplementary ways of favoring the promotion of marginal farmers, in particular by training as many of them as possible and combating land hoarding by minority groups.

Agricultural Challenges to Be Met

These various scenarios have the merit of helping to raise the main questions pertaining to the future of these approaches to Sub-Saharan agriculture. What kind of virtuous circles should be fostered among the determining factors of change to avoid the vicious circles that result in underemployment, famine, conflicts, rural flight, and desertification? How are unproductive family farms going to improve their performance so that they can respond to promising markets without endangering natural capital? How are these farming systems going to increase the share of the value added returned to them within the various agricultural subsectors? What will the balance be between increasing the land area under cultivation and improving productivity? What will be the respective roles of high-performance family farms, agribusinesses, and agroindustries raising crops directly (see box 4.1)? What is to become of the marginalized small farmer families? What does the future hold for young people who would like to set up as farmers and for those who wish to leave the land?

At the current time, efforts to respond to these questions vary widely depending on what responsibilities are held by those expressing themselves. African heads of state tend to focus on the modernization of their agriculture, appealing to a private sector that will invest in agribusinesses capable of tapping the available productive potential better than family farms, which, according to them, will be at pains to increase their productivity. Professional leaders in West and East Africa argue for defending and promoting family farming, for the right to live in the countryside, and for food sovereignty (see part 3). In its 2008 *World Development Report* on agriculture, the World Bank (2007) espouses combating poverty by promoting access to markets, enhancing the competitiveness of small farmers, promoting agricultural and rural employment, improving living conditions in the countryside, and having recourse to migrations and reconversions toward other sectors. The Bill and Melinda Gates Foundation, as part of the Alliance for a Green Revolution in Africa (AGRA), has joined other partners in contributing substantial financing to improve the seeds used by small farmers and to fertilize their land: what remains to be clarified is which seeds are involved and, more generally, how small farmer efforts are to be supported.

Throughout the preparation of this work, formulating the major stakes and challenges to be met has proved to be a complex matter. Ultimately, alongside the challenge of speeding the demographic transition in African rural areas with high birthrates, three challenges appeared as major for African farming systems:

- Tapping a substantial natural potential to accord priority to feeding Africans but also to generate foreign exchange, while ensuring that the food security imperative is not disregarded.
- Promoting the availability of human capital within family farms—which is currently going to waste, owing to the lack of availability of training and innovation—while promoting a favorable social, economic, and political environment.

- Pursuing these efforts to tap natural potential and promote human capital over time, in space, and throughout the entire economy, thanks to the introduction of cohesive agricultural, social, and environmental policies and thanks to integrated land management.

Bibliography

Angé, A. 1991. "La fertilité des sols et les stratégies paysannes de mise en valeur des ressources naturelles. Le mil dans les systèmes de culture du Sud du bassin arachidier sénégalais." In *Savanes d'Afriques, terres fertile?* La Documentation française, Rencontres internationales sur les savanes d'Afrique, December 10–14, 1990. Montpellier.

Bayart, J.-F., H. M'Bembe, and C. Toulabor 1992. *La politique par le bas en Afrique noire: contributions à une problématique de la démocratie*. Karthala, Paris.

Bied-Charreton, M. 2008. "Demain, combien de terres stériles?" Interview by Gaëlle Dupont for *Le Monde,* Janaury 13, 2008.

Billaz, R. 2007. "Développement, emploi et migrations en milieu rural." Seminar on "Les enjeux ruraux et agraires en Afrique." Video on the website of the Gabriel Péri Foundation.

Blein, R., B. G. Soulé, B. Faivre Dupaigre, and B. Yérima. 2008. *Les potentialités agricoles de l'Afrique de l'Ouest* (ECOWAS). FARM, IRAM, ISSALA, LARES.

Brossier, J. 2007. "Apports des théories sur l'exploitation agricole dans une perspective de gestion" *Exploitations agricoles en Afrique de l'Ouest et du Centre,* ed. M. Gafsi, B. Dugué, J.-Y. Jamin, and J. Brossier , QUAE, Paris.

Calame, M. 2008. *La tourmente alimentaire, pour une politique agricole mondiale.* Charles Léopold Mayer, Paris.

Charnoz, O., and J.-M. Severino. 2007. *L'aide publique au développement.* La Découverte, Paris.

Collomb, P. 1999. *Une voie étroite pour la sécurité alimentaire d'ici à 2050.* Economica, Paris.

Deffontaines, J.-P., and J.-P. Prod'homme. 200l. *Territoires et acteurs du développement local, de nouveaux lieux de démocratie.* L'aube, Paris.

Desanker, P. 2007. "Vulnérabilité de l'Afrique subsaharienne." In *Regards sur la terre 2007, dossier énergie et changements climatiques*, ed. P. Jacquet and L. Tubiana. Presses de Sciences Po, Paris.

Devèze, J.-C. 1996. *Le réveil des campagnes africaines.* Karthala, Paris.

———. 2004. "Les agricultures familiales entre survie et mutations." *Afrique contemporaine* 210.

Diamond, J. 2006. *Effondrement, comment les sociétés décident de leur disparition ou de leur survie.* Gallimard, Paris.

Dufumier, M. 2005. *Agriculture et paysanneries des Tiers mondes.* Karthala, Paris.

Dufumier, M., and S. Bainville. 2006. "Le développement agricole du Sud-Mali face au disengagement de l'État." *Afrique contemporaine* 217.

Étienne, G. 2007. "Les dérives de la coopération Nord-Sud, vers la fin des chimères?" *Politique étrangère* 4.

Gafsi, M., P. Dugué, J.-Y. Jamin, and Brossier. 2007. *Exploitations agricoles familiales en Afrique de l'Ouest et du Centre.* QUAE, Paris.

Gigou, J., H. Coulibaly, K.-B. Traore, F. Giraudy, C. O. T.Doucouré, and S. Healy. 2006. "La culture permanente: une innovation paysanne méconnue de la recherche dans le vieux bassin cotonnier de Koutiala (Mali)." *Agronomes et Innovations.* L'Harmattan, Paris.

Giraud, P.-N., and D. Loyer. 2007. "Pour une révolution doublement verte en Afrique." In *Regards sur la terre 2008, dossier biodiversité-nature et développement,* ed. P. Jacquet and L. Tubiana. Presses de Sciences Po, Paris.

Gourou, P. 1991. *L'Afrique tropicale, nain ou géant agricole?* Flammarion, Paris.

Griffon, M. 2007. "Énergie, climat et besoins alimentaires." In *Regards sur la terre 2007, dossier énergie et changements climatiques,* ed. P. Jacquet and L. Tubiana. Presses de Sciences Po, Paris.

Hourcade, J.-C. 2008. "Enjeux géopolitiques du développement durable." *Études* 408 (2).

Jancovici, J.-M. 2002. *L'avenir climatique, quel temps ferons-nous?* Le Seuil, Paris.

Jouve, P. 2006. "Transition agraire: la croissance démographique, une opportunité ou une contrainte?" *Afrique contemporaine* 217.

Kleene, P., B. Sanogo, and G. Viestra. 1989. *À partir de Fonsébougou... Présentation, objectifs et méthodologie du volet Fonsébougou (1917–1987).* Collection Système de production rurale au Mali, Bamako. IER (1), Amsterdam.

Louvel, R., and C. de Gromard. 2007. "L'accès à l'énergie des ressources naturelles." In *Regards sur la terre 2007, dossier énergie et changements climatiques,* ed. P. Jacquet and L. Tubiana. Presses de Sciences Po, Paris.

Mbembé, A. 2000. "Esquisses d'une diplomatie à l'africaine." *Le Monde diplomatique,* (October).

Michailof, S., ed. 2006. *À quoi sert d'aider le Sud?* Economica, Paris.

Meisel, N., and J. Ould Aoudia. 2008. *La bonne gouvernance est-elle une bonne stratégie de développement?* Working Paper 58. AFD, Paris.

Piéri, C. 1989. "Fertilité des terres de savanes. Bilan de trente ans de recherche et de développement agricoles au Sud du Sahara." CIRAD-IRAT.

Raymond, G., and V. Beauval. 1995. "Le coton au Bénin en 1995, situation actuelle et projection à l'horizon 2000 de la production de coton graine." Report for the Ministry of Rural Development.

World Bank. 2007. *World Development Report 2008: Agriculture for Development.* Washington, DC.

Part 2

Steering Transitions of Rural Economies

A review of agricultural and agrarian changes (illustrated in the French version with numerous examples[1]) underscores the importance of the changes under way in both the agricultural and economic systems of rural Africa. But this review begs the question of promoting a transition, not just of agriculture, but of rural economies in general, which explains the title of part 2.

The concept is thus to steer transitions in rural economies as part of societal change. But these terms, of course, deserve further explanation:

- Transition should not be viewed in mimetic fashion as passage from a pre-modern to a modern society, patterned on historical development of the West and Asia. The issue for African countries is to develop in their own way, by seeking to define where they want to go, starting from a well-defined initial situation, and based on the processes of change already under way;
- These processes have to be led by African stakeholders who should specify their objectives to optimize both their strengths and the support of external organizations; these organizations, who long ago began to lose interest in agriculture, need to relearn how to work in this area without preconditions;
- The successful transition of agricultural and rural economies is of critical importance in meeting the growing demand of the market for food products and in providing opportunities to diversify income sources; this concerns the entire rural economy, including nonagricultural activities.
- Given the importance of the demographic, environmental, social, economic, cultural, and political issues, the entirety of rural society is ripe for change.

Chapter 5 stresses the importance of jointly developed government policies that can steer agricultural transitions under the aegis of relegitimized governance. The weakening and loss of legitimacy of national agencies, the persistence of a principle of donor intervention that favors the donor's own particular project approach, and the coexistence of various, more or less contradictory strategy papers are some of the elements of an institutional context that is hardly conducive to the development of such policies. Policy leadership needs to be reestablished at the national level so that Africans can themselves take charge of steering the necessary transitions.

1. "Les trois étapes de la construction d'un mouvement paysan en Afrique de l'Ouest," B. Lecomte; "Innovations et aménagements des bas-fonds en Guinée forestière," J. Delarue; "L'évolution des zones cotonnières de l'Ouest du Burkina," A. Schwartz; "Évolutions des agricultures familiales du Lac Alaotra," J.-C. Devèze; "Les évolutions de l'agriculture malienne sur la période 1970–2000," D. de la Croix; "Le développement agricole post-conflit du Mozambique," F. Desmazières; "Une nouvelle politique agricole au Kenya: nécessaire, mais suffisante?" W. Anseeuw, S. Freguin, and P. Gamba.

The debate over processes, of course, goes hand in hand with the debate over content. The world context has changed substantially in recent years: rising agricultural prices, but rising energy and input costs as well, new environmental requirements, climate change, and more.

Government policies must indeed take this general context into account, but they also need to incorporate answers to the following questions consistent with their specific circumstances:

- How can the imperative of feeding the poor be reconciled with the imperative of producer prices that are sufficiently attractive to encourage growth?
- What innovations should be encouraged in a context of rising energy and input costs and growing urgency to take environmental precautions?
- Given the need to increase production while simultaneously preserving natural capital and fighting against inequalities, what middle ground is possible between expanding the cultivated area and intensification?
- What population and migration policies should be adopted?

The following chapters suggest elements of answers to all these questions and they take a close look at five key areas of government policy: improving and safeguarding systems of land tenure, encouraging innovations and their dissemination, preserving and capturing markets for agricultural products, promoting new financing channels for agriculture, and building the individual and collective capacities of stakeholders.

Interventions in these five areas aim, above all else, to improve the conditions of farm viability, first through the introduction of more effective production systems that are better able to withstand crises of all kinds, but also through improved long-term management of ecosystems and better organization of agricultural sectors.

N.B. The financing of government policies, particularly agricultural policies, is a subject that is not addressed in this book, except tangentially (with references to limited national budgets for agriculture and the decline of official development assistance earmarked for agriculture over the past decade); this topic will assume greater and greater importance as the countries of Sub-Saharan Africa continue to develop, their systems of governance improve, and the opportunities thus increase for implementing appropriate financing tools that may involve direct or indirect taxation.

Also, this book does not discuss the matter of food policies that should round out agricultural policies. Finally, the problem of urban and rural poverty is not examined closely, although this is the leading cause of food problems.

Chapter 5

Agricultural Transitions and the Weight of Government Policies

Vincent Ribier

Transitions in African agriculture must encompass a triad of issues: farm diversification, the interaction between agrarian systems and other sectors, and a secure political and economic environment. This requires a deliberate process of building bridges between the government and all stakeholder categories and, more specifically, efforts to relegitimize and fortify agencies in charge of government policy, strengthen the role of civil society, and refocus development assistance interventions away from a pattern of uncoordinated supply and more toward the principle of demand-driven intervention, involving partners with stronger institutional capacities for meeting the challenges.

Overview of Principal Transitions in African Agriculture

The major problems facing African agriculture prefigure a framework within which African agriculture will need to evolve; by taking a look at these problems, we can demarcate the field of possibilities and describe, from a historical perspective, the agricultural transitions that African agriculture will need to embrace to safeguard the future of African farmers and boost the development of individual countries.

These agricultural transitions are discussed below according to the scale of analysis:

- The first approach to agricultural transition takes place at the farm level. The growing differentiation between types of agriculture has an impact on the corresponding farm units: these become differentiated depending on the evolution of natural conditions and, especially, the socioeconomic conditions bearing upon their activities. Apart from agribusiness operations (with substantial capital and salaried employees), which remain very limited in most African countries, farms are becoming differentiated on the basis of earlier forms of local agriculture. Bélières, Losch, and Bosc (2002) distinguish three such categories: agricultural enterprises, involving the best-equipped family farms or those that have benefited from direct investment in the highest-growth subsectors, a growing fringe of marginalized farms that no longer have the means to ensure their reproduction and that are on a path to impoverishment, and a large middle ground of family farms that find themselves on the razor's edge as a result of market instability or natural disaster. The goal is to understand the trajectory in the evolution of these farm categories by analyzing their capacity for and interest in greater market integration, along with the requirements for transitioning between different types of farms, depending on the context and available support. Among the various factors influencing the trajectory of change, social capital plays an important role. The issue of agricultural transition at the farm level revolves around the conditions that allow family units with fragile or marginalized activities to evolve toward farms that are run like small businesses, capable of creating and selling farm products in line with market demand.

- The second approach to transition involves the agrarian systems. The scale of the agrarian systems is appropriate for dealing with the issue of farm integration within an area[1] and the relationship between those whose livelihood depends on farming and the rest of the population. The scale of the agrarian systems is also appropriate for dealing with the interrelationship between agricultural production and natural resources management, raising questions not only about the more or less aggressive use of natural resources for agriculture but also about the capacity of agriculture to generate environmental advantages. This second approach to agricultural transition, which we could call agrarian transition, is focused on moving away from managing the reproduction of agrarian systems toward managing the evolution of agrarian systems in combination with on-farm changes and changes in the way the natural environment is used.
- The third approach to agricultural transition mobilizes the broadest scale, namely, the political, economic, and regulatory environment surrounding agriculture. In the current state of affairs, this environment is very often unstable, and therefore not conducive to agricultural activity. The government and the stakeholders in agriculture are not always in a position to control this environment, nor to create the conditions to safeguard it. This third approach to agricultural transition focuses on the conditions required for changing the political and economic environment and moving toward land security, effectively managing supply and marketing mechanisms,[2] creating indispensable infrastructure, and making appropriate financial services available. This change requires the best efforts of the government and agricultural trade organizations in fulfilling their responsibilities.

These agricultural transitions must also be placed in the context of transitions in rural society, which call for moving from ever-growing marginalization of poor farmers toward the concept of individuals being able to choose their own future between running a successful farm or succeeding in some other trade. This is the reason government policies not only should be agricultural (market policies, land bureau policies, human and social capacity-building policies, financial policies), but also should take into account the gaps in infrastructure and services frequently encountered in rural areas, urban policies for reducing the poverty of those who are unemployed or who bear the brunt of rising agricultural prices, and demographic policies (see the annex to chapter 1).

Priority Fields of Intervention and Collaborative Development of Government Policies

These three transitions involving farms, agrarian systems, and their economic environment can also be grouped under the heading of "transition of the rural economy" if we add in the growth of multiple activities connected to the non-agricultural rural economy. Success in accomplishing transitions of such broad scope and in such diverse areas is a tremendous challenge; it simultaneously requires the growing involvement of the populations affected; full consideration of positive elements that already exist and alternatives still at the embryonic stage; and speeding up the process of change to ensure that the process does not drag on, which is a potential source of discouragement (Baranski and Robin 2007).

These various requirements for success, taken separately, cannot produce the intended results: only in combination can they do so. Government intervention is clearly desirable to encourage the transitions, accelerate the rate of change, and quickly reach an equilibrium beneficial to all. It is indeed difficult to imagine that such changes can occur spontaneously, based solely on market forces. This deliberate approach cannot achieve its objective, however, without the support of the principal stakeholder groups including, first and foremost, farmers and their organizations. Furthermore, if the relevant stakeholders are mobilized, the objective still cannot be met unless there is validation and support from government authorities. Finally, there is a whole range of embryonic opportunities that cannot truly materialize without a nudge in the right direction.

The combination of a deliberate approach and substantial stakeholder involvement defines the general participatory framework in which interventions must be developed and implemented. The challenge is to create favorable institutional conditions for genuinely collaborative policies to emerge. This raises the issue of how to develop policies rooted in a shared vision of the future of agriculture and of the role that each stakeholder will play. Producer organizations and, more generally, the various forces of civil society have a decisive role to play in developing and defending a vision of the future of agriculture involving broad grassroots participation. In concrete terms, development of this shared vision of the future of agriculture, and especially of family farms, hinges on success in working out compromises in a number of areas such as customs protections, the place of agribusiness, the scope and points of application of farm subsidies, and land tenure regulations.

The Need for a Change in Policy Governance

The current paradox is that this need for government policies is being felt right at the time when governments are especially vulnerable and their room to maneuver, from a financial perspective, is very limited. A quick look at the way in which the government policy process plays out is sufficient to confirm their vulnerability and, in particular, their dependence on external forces, both of which undermine any expression of national leadership.

An Institutional Situation That Is Greatly Eroded and Thus Hardly Conducive to the Development of Government Policies that Are Capable of Addressing the Problems Raised

Loss of Legitimacy of National Agencies and Diminishment of their Capacity to Steer and Coordinate the Policy Process. The causes of the erosion of the institutional situation lie first of all in the weakening and loss of legitimacy of national agencies responsible for steering agricultural policy. The various structural adjustment programs (SAPs) implemented in the 1980s led governments to reduce the financial and human resources of their agencies. The SAPs in fact resulted in the departure of many civil servants; those who remain are typically older (the age pyramid for most Ministries of Agriculture in Africa shows that the age bracket of 50–60 years is overrepresented in relation to the age brackets of 20–30 and 30–40 years), and the low wages for the younger workers push them to seek additional income outside their principal activity.

The institutional and political crises experienced in many countries over the past two decades have also played a role in paralyzing the operations of national agencies for more or less extended periods of time (roughly 1 year in Madagascar, 15 years in the Democratic Republic of Congo). They have resulted in frequent ministry reshufflings that have impaired the effectiveness of services and prevented clear mandates from being given to each governmental unit, thus encouraging rivalries and overlapping activities.

Implementation (Financing) of Interventions That Are too Dependent on Donors Acting in a Disorganized Manner. This weakening of national agencies has continued over the past decade as a result of frequent distrust on the part of donors. A large majority of donors, judging national government agencies to be ineffective, have preferred to turn to private actors or civil society to implement their development assistance programs. This strategy, adopted on the grounds of government inefficiency, in fact serves to weaken even further the already anemic agencies. The example of agricultural statistics in Madagascar is particularly instructive. Taking note of the inadequacies of the Malagasy agricultural statistics system, the European Union has undertaken a data collection and processing program that is run by a Belgian nongovernmental organization (NGO), instead of supporting the agricultural statistics service at the Ministry of Agriculture. Such a system clearly does not strengthen the latter, which instead loses all legitimacy: both national and international partners are going to rely directly on the Belgian NGO, which has now become the sole credible spokesperson on agricultural statistics. As a general rule, the consequences of setting up ad hoc structures not directly tied to government agencies can be felt on two different levels:

- First, the governmental units that are supposed to be in charge of projects and to ensure the consistency of the national strategy find themselves out of the loop and deprived of their responsibilities; they have even fewer resources than before and, to a large extent, lose all credibility in the eyes of stakeholders.
- Second, the development of employment opportunities that pay much better than the civil service causes some of the most highly trained officials to leave government service to seize such opportunities. In Madagascar, which is hardly an exception in this regard, the wages of Malagasy officials working in NGOs are roughly four times higher than what they could earn at the Ministry of Agriculture!

An Abundance of Strategies Undermines the Strategy. The weakening of national agencies and the decisive influence of uncoordinated donors result in a proliferation of agricultural development strategy papers that are not always consistent among themselves. Thus, at the same time, there may well be a variety of different documents specifically targeting the agricultural or rural sector, including letters of agricultural policy, rural development guidelines, strategy papers for the rural sector, and both master plans and business plans. The different documents are generally produced by different entities (different directorates of the Ministry of Agriculture, the Ministry of Planning, the Ministry of Finance, and, in some cases, the Office of the Prime Minister or even directly the Office of the President), supported by different donors, which explains the coexistence of a range of documents with essentially the same objectives. The issue of superimposition is also thematic in nature: indeed, one finds a proliferation of strategy papers in which the central theme is different but the areas of intervention overlap. In addition to documents that focus strictly on development of the agricultural sector, there are documents targeting themes that have gradually become impossible to ignore, such as sustainable development, decentralization, institutional development, and poverty reduction.

The coexistence of documents that address the same development issues but from different angles and without any real harmonization often reflects the lack of a consensus on a true development strategy. In general, there are two distinct trends. One trend promotes agricultural development on the basis of agricultural entrepreneurship as a way to modernize family farms. This means that only the most dynamic will survive, while the others will have to adapt, migrate, or vanish. This perspective is set forth in master plans or business plans. The other trend favors comprehensive development that incorporates both environmental concerns and issues of fairness. This vision seeks to avoid further marginalization of small farmers and instead provide them with development alternatives, and it thus entails a participatory approach to the policy development process. The coexistence of different and distinct orientations maintains a degree of ambiguity and shows that there is really no clear arbitration on the part of governments in favor of one model over another.

The Need for a Recovery of Policy Leadership at the Country Level

The institutional context described above clearly hinders the progress of the various agricultural transitions. In this regard, a change in policy governance is needed, one that should be rooted particularly in relegitimization of national agencies in their coordination role and heightened involvement of producer organizations and the forces of civil society, along with changes in official donor intervention arrangements.

Relegitimization of National Agencies in their Coordination Role. National agencies are today too lacking in legitimacy and too weak to play a role in coordinating new-generation government policies. Yet, they are indispensable in this role. The debate on the proper role of government has not been immune to the vicissitudes of fashion. In the 1960s and 1970s, government intervention was deemed legitimate in and of itself, and it would not have occurred to anyone to try to justify it. The liberal wave of the 1980s and 1990s then swept away these old certainties to tout the supremacy of the marketplace. Today, retreating from these new excesses, the debate is characterized by a pragmatic search for complementarities between the market, the government, and other public and private stakeholders active in the sector. The role of government needs to be rewritten on a case-by-case basis with all the other types of stakeholders. The recent publication of the World Bank's *World Development Report 2008: Agriculture for Development* is fully consistent with this evolution.

Opening up of the Political Spectrum, Strengthening of Producer Organizations, and Decentralization: The Strength of Civil Society. Public policies of the 21st century will not really be public unless there is greater participation from the various members of civil society, particularly farmers. The vitality of the different stakeholder categories and their capacity to become involved vary greatly from one country to the next. Overall, producer organizations and other forms of organizations are growing stronger in many countries of West Africa, which is less true in Central Africa. To a large extent, this hinges on the openness of the political spectrum and the scope of the decentralization initiatives undertaken by various governments. The institutional fabric is beginning to change and grow richer, all the more so when government plays its role in safeguarding the socioeconomic environment, supporting stakeholder dynamics, and developing projects that promote the common good. The various forces of civil society have undergone gradual development:

- Local governments and other forms of local authorities (districts, councils, assemblies) have assumed greater importance, at least in some countries, as a result of the decentralization process. The leaders of these entities draw their legitimacy from their status as elected officials and they are not afraid to oppose agencies under whose authority they do not fall. Accordingly, they may constitute opposition forces in countries where the political spectrum has been opened up. The difficulty in clarifying the responsibilities of local governments, and the scant resources at their disposal, complicate the institutional landscape and may cause growing confusion between different decision-making levels, from the subnational level (local communities) all the way through the supranational level (international executives).
- At the same time, producer organizations and, to a lesser extent, joint trade organizations have gradually become stronger and therefore play a growing role in the agricultural sector. Virtually nonexistent in the 1970s and 1980s, producer organizations began to develop in the 1990s on a number of levels: local (village groups), national (such as the National Council for Rural Consultation in Senegal and the National Union of Cotton Producers in Burkina Faso), and subregional (West African Network of Farmer and Producer Organizations). Three categories of local organizations have been recognized, the specific features of which are sometimes fluid in the field: specialized groups related to a particular industry and that perform specific economic functions, multisectoral groups that attempt to accommodate the diversity of activities of their members, and) organizations for specific social categories (women, youth).

- Nongovernmental organizations have proliferated over the past 20 years in most African countries, benefiting from the growth in opportunities to receive donor financing. Accordingly, they maintain a large presence in all countries as development project operators. However, NGOs are a very heterogeneous group in terms of size, areas of intervention, and the objectives they pursue. Some are purely national, while others are field units of international NGOs. Some pursue medium- and long-term development goals, while others are more opportunistic and operate essentially as consulting firms in search of financing.
- Private operators, from small local businessmen and entrepreneurs to large private operators affiliated with multinational firms, are also important stakeholders in the agricultural sector. They have to some extent emerged as a substitute for government agencies in the processing and marketing of products. Their power to exert pressure has risen substantially, and they are in a position to wield considerable influence over government decisions.

Donor Changes and Adaptations

Donors hold decisive influence in the institutional landscape of African countries. They greatly influence development strategies in general and agricultural policies in particular. The region's traditional donors (World Bank, European Union, EU member states, and especially France in the francophone countries) have recently been joined by new donors who are often more pragmatic and less demanding in terms of conditionalities, such as China, Japan, the Russian Federation, and the Arab Funds.[3] This diversification of potential sources of financing clearly strengthens the government's negotiating capacity, but it does not necessarily create the right conditions for ensuring that interventions are consistent with a national strategy. As noted by Bergamaschi (2007), donors' attitudes are often ambiguous: they tout the importance of good governance and the need for government to take ownership of policies, but their actions in some cases belie these goals. Many donor-financed institution-building and capacity-building programs fail to achieve their expected impact in terms of building stronger national structures. The frequent tendency to create ad hoc units, possessing better working conditions than usual, but with no guarantee of longevity, is indeed a way to get the structures up and running quickly and to tailor institutions in line with donor requirements, but only too rarely does this create the conditions for lasting institutional development.[4]

The donor track record in terms of support for particular policies is very mixed. Official development assistance appears to have created some degree of dependence on the part of governments and, overall, capacity-building interventions have not achieved their intended effects in terms of institutional reinforcement. Donors have, in most cases, clung to supply-based principles by proposing interventions in a disorganized manner, while governments have for the most part been unable to formulate institution-building strategies that could serve to channel such interventions. Rivalries between units and individuals within government agencies, on the one hand, and the shortage of interministerial coordination, on the other, have exacerbated competition between government entities and led many of them to accept projects that were not mutually consistent or that failed to match national development goals.

The procedures and arrangements governing assistance clearly call for improvement. A brief survey, not meant to be comprehensive, is instructive in this regard:

- Advocacy for institutional support for national dynamics conducive to policy ownership. The main thrust of policy support should be to contribute to the emergence of a national debate, enhance the collective capacity to identify and address the principal policy challenges, and support the dynamics for building concerted government policies. The general objective is to strengthen the national capacity to develop and implement policies that match a collectively selected development model.
- Support for the emergence of government policies involving all the different categories of public and private stakeholders. The emergence of strong government policies requires the participation of social stakeholders in the sector, particularly collective organizations representing those who, so far, have typically been largely excluded from the design and potential benefits of such policies. The goal is to promote a policy comanagement model in which the government has a legitimate role in its function as guardian of the public interest and in the steering and coordination of interventions, on the one hand, while the various stakeholders (producer organizations, private sector, consumers, local governments, national representatives) are directly involved in the debate and the arbitration process, on the other.
- Institutional support for government agencies is a function that should be maintained and even intensified inasmuch as it bears special relevance in the current context of redefining the role of each stakeholder category at the national level and opening the commercial parameters. The capacity of the Ministries of Agriculture to justify and argue their choices and programs (to other ministries, private partners, and donors) needs to be strengthened. The government needs to be relegitimized in regard to its role in steering the agricultural policy process, and its capacity to steer this process needs to be enhanced. The challenge is all the greater because governments have been durably weakened in past years, and many current donor interventions do not help strengthen governments in their role of steering and coordinating resources and interventions that target the agricultural sector.

- Building the government's steering capacity goes hand in hand with promoting a dialogue between government and stakeholders (producer organizations, private sector, civil society, and so on), which has not yet become standard practice. There has indeed been progress in recognizing the place and role of nongovernmental stakeholders in policy development,[5] but many civil servants remain reluctant to share their old prerogatives. Major efforts should therefore be undertaken to overcome this reluctance and to promote the dynamics for cooperation and collaborative development of agricultural policies. Support for different stakeholder groups presupposes a realistic view of the social relations at play, especially with respect to the reduction of poverty and inequities: advancement of the middle classes under the guise of defending the underprivileged may turn out to have no driving effect on improving the living conditions of the latter. However, broader participation by a large number of stakeholders remains the best way to guarantee the adaptability, realism, and effectiveness of interventions. It allows the different stakeholders to be informed, to assess any conflicts of interest, and to accept the compromises needed between different scales of contemplated activities.

Steering agricultural transitions on the part of those most directly involved requires a government policy process that escapes the constraints of the existing institutional landscape. This significant challenge is quite clearly far from being resolved, but in many countries there are encouraging signs, specifically in regard to the will of government to reassert policy ownership. This willingness is most clearly apparent in the implementation of participatory processes for developing laws on agricultural policy, as in Senegal and Mali, in a clear break with the earlier usual practices. Producer organizations have gradually increased their capacity to weigh in on the agricultural policy debate in various countries. They carry the dual requirement of clarification of agricultural development options and assurance of appropriate resources for turning these options into concrete action.

Notes

1. The discussion here does not address the living conditions of farmers and their families, which are so important to their future, nor the matter of multiple activities that can offer a greater range of opportunities for supplementing income.
2. The principle of a market-oriented transition that is largely embraced by most African family farms has not been advanced, as outlined in the World Bank's *World Development Report 2008*.
3. There are no fewer than 40 different donors in Mali, the largest of which are the European Union, France, and the World Bank, followed at a considerable distance by the Netherlands, Japan, the United States, the African Development Bank, Canada, and Germany.

4. The main interventions targeting national government agencies have consisted of programs to support the reform and restructuring of these institutions. Although the goal of such programs has been to reform national agencies so that they will be in a position to handle new functions as part of the liberalization process, in practice these programs have had the effect of reducing civil servant staffing levels, without however, permitting the rehiring of more qualified staff, or capacity building for those remaining in the civil service. The Agricultural Services component of the PASAOPs (Agricultural Services and Producer Organization Support Projects backed by the World Bank in various countries) has encouraged the creation of parastatal agencies in lieu and in place of government research and extension services. As a result, government agencies have been relieved of their public service mission, in favor of entities that hold an uncertain future and whose survival depends on their ability to get solvent producers to pay for their services.
5. A number of institutional support projects of the French Ministry of Foreign Affairs have made efforts in this direction; in Vietnam, for example, this approach resulted in a change in the demands of the Vietnamese Government (popular committee), initially technical in nature, toward more methodological and pedagogical concerns, with a growing interest in participatory approaches.

Bibliography

Baranski L., and J. Robin. 2007. *L'urgence de la métamorphose.* Des idées et des hommes, Paris.

Bélières J. F., B. Losch, and P. M. Bosc. 2002. "Quel avenir pour les agricultures familiales d'Afrique de l'Ouest dans un contexte libéralisé?" IIED, London.

Bergamaschi, I. 2007. "Mali: Patterns and Limits of Donor-Driven Ownership." Managing Aid Dependency Project GEG Working Paper, 2007/31. Oxford University.

Chabassou A., and M. Ruello. 2006. "Etude d'un processus de concertation pour l'élaboration d'une politique publique: le cas de la Loi d'orientation agro-sylvo-pastorale (LOASP) sénégalaise. " CIRAD-ES Working Paper. Paris (December).

Chedanne, Ph. 2005. "La crise des politiques agricoles en Afrique." *Lettre des économistes de l'AFD* 10 (December): 2–6.

Félix A. 2006. "Eléments pour une refonte des politiques agricoles en Afrique Subsaharienne." *Afrique Contemporaine* 217.

Ribier V., and J. F. Le Coq. 2007. "Renforcer les politiques publiques en Afrique de l'Ouest et du Centre: pourquoi et comment?" *Notes et études économiques* 28: 45—73 (Ministry of Agriculture and Fisheries).

World Bank. 2007. *World Development Report 2008: Agriculture for Development.* Washington, DC.

Chapter 6

Land Policy: A Linchpin of Economic Development and Social Peace

Philippe Lavigne Delville and Vatché Papazian

What land policies can safeguard farming activities, and also boost labor productivity, avoid growing inequalities and tensions, and better manage natural resources over time? Given the complex historical heritage, the diversity of ecosystems, and the growing appetite for land, to imagine and actually implement new policies requires mobilizing the support of stakeholders (particularly the government and producer organizations) and raising the necessary resources. One issue that deserves attention is the evolution of farm structure and especially farm size.

The question of land policy encompasses the full breadth of relations among and between stakeholders in regard to land and natural resources (Le Bris, Le Roy, and Mathieu 1991).[1] The manner in which a society defines ownership and user rights to land and natural resources reflects the conception that members of this society hold concerning relations between individuals and between people and nature.[2] Land policy thus lies at the heart of the economic, social, and political challenges facing a society; land is the basis of both farm and pastoral production, and the distribution of rights to land and resources typically reflects social and economic inequalities. Furthermore, many conflicts, whether local or international, involve land disputes (at least in part), either because of competition for the land or because the land issue is being exploited for other purposes.

The importance of land policy is all the more critical at a time when social and economic inequalities are on the rise, the world is becoming increasingly urbanized,[3] new needs stemming from demographic and economic change are threatening to marginalize a whole swath of the farm population, and the urgency of environmental challenges is greater than ever. By defining legally recognized land rights, influencing the distribution of rights between stakeholders, and specifying how these rights are to be managed and transferred, land policies affect the very foundations of society.

Since the issue of land policy lies at the crossroads of demographic, economic, environmental, and social concerns—all the more so when pressure on the land is increasing—and because the sustainability of natural resources is at risk, the question of land policy permeates this entire book.

This chapter attempts to assess the diversity of situations and policies that currently exist, lays out the major land-related challenges facing rural areas, and then suggests possible approaches along three complementary dimensions: first, land policies for safeguarding agricultural activities; second, structural improvements for farms; and third, the management of natural resources and local lands for sustainable agriculture.

Although land issues are global in nature and relevant to all situations, this chapter focuses specifically on Sub-Saharan Africa. No region has ever before had to manage such a high population growth rate and such increased upheaval stemming from the land use and resource management practices of societies and governments.

An Increasingly Worrisome Land Situation

In Sub-Saharan Africa, a great variety of land situations are found, ranging from arid zones to equatorial forests, and from largely unpopulated spaces to saturated villages and periurban areas. Elements of diversity include natural habitats, population densities, ways of using the environment, local methods of control over land and resources, land policies and natural resource management policies, existence or nonexistence of farm water systems, influence of urban stakeholders and agricultural entrepreneurs, government decentralization policies, and more.

Current Land Policies and Background

The colonial chapter of history left its mark. Southern Africa (and, to a lesser degree, East Africa) experienced heavy agricultural colonization, resulting in the creation of a dual agricultural system, split between large "white" modernized farms and small "black" farms on the less desirable land. This duality endures despite the accession to power of national elites who have often sought to recover the large farms. The issue of land reform remains a live topic in some countries, notwithstanding recognition that the difference in the level of technology between these two forms of agriculture makes simple redistribution an unlikely prospect. Elsewhere, in places where colonial laws had relatively little effect on local land use practices, the bulk of agricultural production comes from family farms, modernized to a greater or lesser extent, based on local land rights, most often with no official legal recognition.

The newly independent governments perpetuated, virtually intact, the land prerogatives instituted by the former colonial powers. They retained the principle of "*domanialité*" in the francophone countries and of "trustee" in the anglophone countries, with an eye to the consolidation of young governments and modernization, sometimes under the authority of a socialist regime. In some cases they even tightened the statutes from the late colonial era, further undermining local rights. Even if the land issue did not seem to be of critical importance, the government wanted to be able to mobilize land for its modernization schemes—infrastructure, migration-dependent development projects, and the like—yet this served only to prolong the legal duality and encourage land grabs by the urban population.

Some governments wanted, more or less transparently, to attempt the socialization of production, sometimes in the form of state farms or cooperatives (such as oil palm plantations in Benin), and sometimes in the form of village groups of family farms (such as Tanzanian "villagization"). In Zimbabwe, the brutal expropriation of colonists has resulted in recent years in a hasty redistribution of land to unprepared populations, thus creating a production crisis.

With respect to natural resources, political and government officials retained and sometimes strengthened exclusionary policies under the guise of "rational" management: reaffirmation of the state monopoly over ligneous resources, distribution of operating licenses to the regime's allies, nature conservation policies excluding local inhabitants, and so on (Compagnon and Constantin 2000). Some governments placed major portions of their country in reserves, either for tourism or hunting, going so far in certain cases as to destroy the local population or condemn it to famine. Fully perpetuating the culture of forest officers of the colonial metropolises, the forest services excluded farmers from the exploitation of ligneous resources, with rules placing them in permanent violation. In a similar vein, efforts were sometimes made to "rationalize" transhumant stockraising, by pushing to make pastoralists sedentary, in contradiction to the random nature of pastoral resources in arid areas.

As a further legacy of colonization, heightened by a government vision of how to modernize backward rural populations, this clash between the technocrats' perspective and the principles guiding farmers' decisions has greatly impacted sectoral policies on natural resources.

As a result, land policies in rural areas are characterized by substantial ambivalence. On the one hand, the primacy of private property, based on a title, is reaffirmed but without anything being done that would enable rural inhabitants to become property owners. On the other hand, the great majority of lands fall, in theory, under state authority, and rural inhabitants cannot, in practice, obtain legal recognition of their land rights. While some areas have been targeted for major interventions (such as infrastructure development or organized migrations), laissez-faire policies have been largely dominant in all areas that the government does not consider to be of major economic or political importance, and land legislation has remained largely or wholly unenforced. Consequently, instead of land policies per se, it is more specifically the land settlement policies, the policies to promote cash crops (such as cotton, groundnuts, coffee, and cocoa), and the farm water systems development policies that have had the greatest impact on land distribution and that have shaped local property systems. Through this diversity of situations, a "right of practice" has emerged (Hesseling and Le Roy 1990) based on customary foundations, transformed to a lesser or greater extent by social and economic change and government intervention.

Furthermore, because the economic and political stakes of land issues are so high, the land service is often a prime locus of corruption and patronage. Because of problems at the land service, some individuals—even if they hold a legal title—may find themselves in an insecure position: land that was not legitimately obtained from the landholder, social pressure that makes it impossible to farm the land, the coexistence of several titles for the same piece of property, and so on.

Growing Tensions and Conflicts

In rural areas, the rights of farmers over the land they work have often, in practice, been consolidated as land has gradually been assigned and cleared, with the customary authorities then playing a regulatory and conflict management role. In highly hierarchical societies, however, the customary authorities continue in some cases to claim the right to assign farmlands and pasturelands, thus keeping farmers in a precarious situation, sometimes with the complicity of the central power as it seeks to form alliances with local chiefs. This is the case in Ghana, for example, and in northern Cameroon (Gonné and Seignobos 2006). Areas experiencing rapid land saturation are sources of tension and form a breeding ground for land claims based on indigenous status or exclusion of "foreigners"; these tensions are often exacerbated by politicians.

In addition, political and social upheavals and the economic marginalization of entire territories encourage assorted reactions: revolt movements, exclusion of emigrants, reconversion of marginalized spaces to grow illicit crops, and armed movements settling into marginalized areas. Ultimately, urban and rural inhabitants who possess a relationship with the powers in place manage to assemble larger or better developed farms, or both; drawing on their financial capacities and the laws in effect, they invest in land in anticipation of urban expansion to outlying areas, or to prepare for their retirement by creating plantations, or to diversify their income stream by venturing into farm crops. They sometimes use legislation to grab land at the farmers' expense, and they sometimes purchase land at a relatively low price from farmers in distress or from local authorities seduced by the lure of gain.

Population growth and agrarian crises in arid or saturated areas feed rural migrations toward virgin or less populated lands. Pioneer settlements in tropical forests—with fragile soils, and that play a critical role in regulating the world's climate—are currently much more extensive in Latin America than in Africa. In contrast, there have been many migrations in Africa into agricultural and stockraising areas that are sparsely populated or more appealing; some have been organized by governments (around the Volta River valleys in Burkina Faso and the Bénoué Valley in Cameroon) or else encouraged by them (for example, into the Ivorian forest region for coffee and cocoa, and western Burkina Faso for cotton). Some regions that have experienced heavy immigration now experience strong land tensions and are highly sensitive to political and ethnic manipulation of such tensions (Côte d'Ivoire, Ghana, Kenya). Seeing the local lands saturated, young indigenous inhabitants seek to challenge the agreements by which their parents granted land to migrants for the purpose of "settling," with no time limit specified. Sometimes, they try to recover land allocated long before; in other cases, relatives sometimes attempt to obtain income from the migrants by leasing the land to them, as in the cocoa-producing area of Ghana, resulting in intrafamily and intergenerational tensions (Amanor 2005).

Land pressure also leads to conflicts over the use of natural resources. Thus, farmers seek to increase the amount of land under cultivation, while pastoralists try to preserve their pasturelands. When livestock corridors and access trails to water points are placed under cultivation blocking herd movements, then pastoralists are likely to cross cultivated fields, causing damage that is a source of conflict. Furthermore, the number of animals kept by farmers is growing, and they also need pasturage, resulting in increased livestock pressure. Last, rural and urban inhabitants also compete to control firewood, the leading energy source for cooking.

Land policies are sensitive in nature, entail important economic and political dimensions, and are thus readily exploited in political conflicts. The growing number of armed conflicts that are linked to some extent to land issues tend to disrupt existing land practices and create new tensions, given the pressure placed on natural resources by displaced populations. Land issues constitute one of the dimensions of the conflict in Côte d'Ivoire, for example, and in Darfur. In the case of the Rwandan genocide, the small farmers' need for land was no doubt an aggravating factor that contributed to the scope of the tragedy, even if this cannot be considered its source.

Whether between migrants and the indigenous population, between urban and rural inhabitants, between farmers and pastoralists, or within the same family concerning access to land, conflicts that are linked to land issues or that have a land dimension are proliferating in rural areas. They are sometimes the result of competition for space, in areas where land pressure is intensifying. They may also stem from contradictions between different rules when the law conflicts with the principles underlying access to land and natural resources for farmers and pastoralists, with some stakeholders more aware than others of how to take advantage of the multiplicity of standards. These conflicts are also the result of social and economic change, but they are linked as well to the fact that, sometimes, neither customary practice nor the law provides answers to new questions that arise. Can land be sold? Under what conditions, and by what procedures? How can the conditions for complementarity between farming and stockraising be reestablished when the supply of available land is disappearing? What rules should govern access to parcels when bottomland is developed, or to water when a water hole is improved for passing herds? And what rules should govern access to land in areas that have witnessed heavy migrations when the rules that were set by the fathers no longer work well because the space has been saturated? How can viable farms be maintained despite this heritage?

Given all these questions, the challenge is not just a matter of formalizing land rights and the land service; first and foremost, it is a governance issue. What are the rules for meeting these challenges? What are the appropriate institutional mechanisms for setting the rules and seeing to their enforcement, in a context where government mechanisms and customary organizations are too often in conflict or ignore each other?

The Major Challenges of Land Policy Reform

In this context, the major challenges for devising appropriate land policies that encourage economic and social development in rural areas and promote social peace can be grouped under five main themes.

Take into Account the Multiplicity of Local Rights and Standards

Because of the region's history and the incomplete social and political integration within countries, most societies in countries of the South are composite in nature, bearing substantial social and cultural heterogeneity; they bring together different social groups (local, so-called traditional societies, farmers integrated to a greater or lesser extent in world markets, urban middle classes, inhabitants of city outskirts and slums) that have different visions of the world and different systems of reference.

The social standards and the conceptions of land policy held by these different social groups are varied and may change over time. In rural areas, these standards constitute the foundations for exploiting the environment that have evolved over the course of history. Thus national laws are superimposed on a variety of standards and methods of land regulation, which they then transform to a lesser or greater extent. Too often, they are rooted in principles that fail to take into account major elements of the social realities and are organized in such a way as to exclude large segments of the population from access to their rights. Sometimes they seek to impose on the entire country a standard derived from a given social group or a particular religious reference. The underpinnings for local conceptions of land policy are often ignored, regardless of whether they are based on a population's traditions and extended history (customary systems) or on recent history (rules instituted in connection with the development of farm water systems), even though, in practice, land regulations hinge on these underpinnings in large parts of the country.

This disconnect between legality, legitimacy, and practices keeps a major portion of the population in an extralegal situation. It provokes, to varying degrees, conflict and opportunism. How best to take into account the diversity of land standards, what status to give to local rights, and how to process the duality between national laws and local standards are some of the main challenges. Shedding archaic vestiges of the colonial past, recognizing the diverse sources of rights so as to organize them more effectively, and basing legal recognition of rights on legitimate and peaceful uses are all imperative.

Manage Land Disputes and Reduce the Sources of Conflict

Four main types of land dispute can be identified, as described below (Merlet 2002).

- *Conflicts over the distribution of access rights to land or renewable resources* (forests, water points, pasturelands, etc.). In some cases, the conflict is the result of great inequalities, rooted in history, such as the conflicts between large landowners and marginalized small farmers. In other cases, the conflict is caused by a land grab process, resulting in the exclusion of stakeholders who are already present and settled, sometimes at the government's initiative (for example, when a protected area is created or a dam is built), and sometimes as a result of market forces (for example, urban land grabs for speculative purposes in periurban areas or spontaneous settlements of migrants in sparsely populated areas).
- *Conflicts in regulating the coexistence of different uses of the same space*: between farmers and pastoralists; between foresters and farmers; and between urban, industrial, and agricultural uses of water resources.
- *Conflicts involving a lack of secure rights and the absence of legal recognition* (customary rights and the rights of farmers, sharecroppers, and tenants).
- *Conflicts over territorial control and the defense of identities.* This concerns, in particular, local, historically autonomous societies that see external actors infringing on their territory, either for mining or forestry purposes or to clear and cultivate the land. Local identities are often tightened as a

corollary, since advancing a connection—real or assumed—between territory and socioethnic identity is a way to legitimize the political claim. Apart from areas of industrial, mining, or forestry operations, this type of conflict is also found in areas that witness heavy agricultural migrations, when land pressure and a sentiment of being dispossessed generate or encourage claims of indigenous rights and delegitimization of the right of migrants to hold land.

These conflicts encompass, to varying degrees, economic dimensions (access to land, to means of production, and to the foundations of subsistence), social and identity-based dimensions (between groups that define themselves as being in opposition over the stakes in question), and political dimensions (distribution of resources between social groups, territorial control, and imposition of rules and arbitration). If the conflict is inherent in social life, then the fact that land disputes may crystallize and take a violent turn, or even assume a regional dimension, stems partly from the shortcomings of land policies and partly from inadequate methods of arbitration and regulation, regardless of whether these methods are political or judicial, formal or informal.

In a growing number of rural areas, the task of preventing and settling such conflicts poses a major challenge. Apart from legal modifications that reduce the extent of legal duality, there is also a need to promote local processes for negotiating and consolidating legitimate rules to address this challenge and to regulate competition between stakeholders, which means that conflict resolution and arbitration mechanisms need to be strengthened. In addition, the government must play its role in establishing a framework that provides straightforward and reliable solutions and ensures compliance with agreements (Gonné and Seignobos 2006).

Safeguard Land Rights and Land Transactions

The security of land tenure refers to the right of all individuals and groups to real protection of their land rights against attempts by third parties or forced evictions. People cannot make investments unless they are assured of the opportunity to benefit from the fruits of their efforts. Land security is thus a requirement for economic development. Sometimes because their social position is not guaranteed under customary systems, but more often because they do not have access to legal recognition of their land rights, many people find themselves in a situation in which they lack land security. Based on recognition of the diversity of rights and sources of legitimacy, ways to safeguard this diversity of rights need to be devised. This requires reliable conflict management mechanisms, along with the formalization and recording of some of these rights.

Inasmuch as rural Africa remains in a state of "incomplete marketization" (Le Roy 1997), one can rarely speak of a real estate market. However, in many regions, market transactions involving land are growing more common. They are sometimes illegitimate and clandestine. But sometimes they are now socially accepted. A large proportion of conflicts between farmers involve sales, either because of ambiguity or manipulation regarding the content of what is being sold (the land itself or cropping rights) or because the procedures for ensuring that a sale is both legitimate and legal have not been stabilized. How can the consent of eligible family members be guaranteed, and how is it possible to avoid a situation in which a sale arranged by the head of the family is contested by others? How can double selling be avoided? In this regard there is a clear need for government intervention to provide straightforward, reliable, and operative mechanisms.

This raises serious issues for policies aimed at promoting land security. Everyone recognizes that, to produce, farmers and pastoralists need to be able to benefit from the fruits of their efforts, and that their access to land and resources therefore needs to be safeguarded. But what constitutes optimal land security? And, more fundamentally, who provides this security? When land pressure is great, is it possible to provide such security to everyone at the same time? Can policies to promote land security choose to overlook a political decision concerning the type of farm to be favored? More broadly, how are these real estate markets to be regulated? Policies to promote land security must be rooted in the recognition of rights to land and natural resources, in all their diversity, and not based on a single model.

Reduce Inequalities and Exclusions in Access to Land

In some countries or regions, access to land is profoundly unequal. Such inequalities—excluding a large part of the population from access to the means of subsistence—force individuals to be employees or sharecroppers of the wealthiest farmers or agricultural enterprises, typically in an asymmetrical relationship. Moreover, the wealthiest farmers are not always the most dynamic. A drastic reduction of the inequalities in access to land is urgently needed in these countries, notwithstanding the fact that deliberate agrarian reforms, through expropriation or "market-assisted," are delicate to implement.

Farmers frequently leave land to their children who wish to live separately, but these are not necessarily the best parcels of land and their children do not always possess the necessary means of production. Women often have a hard time obtaining land, as shown by examples in Burkina Faso, where women are often left only small fields with the most degraded soils (Konaté 2006). For both women and youth, who together account for a major portion of farm output, access to land is becoming problematical.

Apart from agrarian reforms, there are a number of possible ways to promote a socially desirable distribution of land that combines fairness with economic effectiveness. These include regulating land markets, instituting a tax system that discourages unproductive uses of the land, and promoting credit for buying land.

Promote Sustainable Management of Rural Ecosystems

In rural areas, environmental concerns revolve first and foremost around the issue of sustainable management of ecosystems. The challenges are many and their acuteness varies, depending on ecosystem and climate, but they include the need to regulate land clearing and preserve forests and pasturelands; avoid the cultivation of lands unsuitable for farming; practice appropriate cropping techniques to avoid soil degradation; manage surface waters to minimize the risks of erosion, flooding, or drought; permit replenishment of the groundwater table and avoid overexploitation for irrigation; promote the retention of organic matter in soils; reduce the use of fertilizers and crop protection products in areas where they pose environmental problems; and control desertification. All of this requires a multipronged approach that includes physical improvements to the environment, a change in cropping practices, regulation of these practices, and farming-stockraising links.

Among other things, sustainable management of natural resources requires negotiating or renegotiating access rules and user rights in light of the current challenges. Based on values and criteria that make sense to stakeholders, the issue is to negotiate the rules and set in place regulatory mechanisms to ensure that exploitation does not extend beyond natural regeneration. This means giving exclusive rights to local communities and setting rules of access for third parties, depending on the status of the resource (box 6.1).

Land and Sectoral Policies for Strengthening Family Farms

In rural areas, depending on the population density, the ecosystems, the production systems, and the economic environment, population growth may result in increased land pressure. How can rural spaces be settled more densely without worsening the phenomenon of rural poverty or degrading the ecosystems?

In Africa and elsewhere, family farms are the form of agriculture that optimally combines economic performance and income distribution. Given the ecological and economic constraints weighing upon them, along with the fact that they have hardly benefited, with a few exceptions, from supportive policies, their performance is quite remarkable. Yet they still fall well short of the challenges that present themselves.

A true agricultural revolution is necessary if family farms are going to be able to meet the requirements of diet and the production of agricultural or pastoral

BOX 6.1

Communal Management of Land Rights (Benin, Madagascar)

Exploring innovative ways of achieving land security is one goal of current land reforms. In parallel fashion, Benin and Madagascar have each developed policies to ensure land security based on analogous principles:

- Move away from legal duality and give legal recognition to land rights acknowledged by consensus at the local level. This involves drawing up "land certificates" and creating a new legal status for "rights established or acquired in accordance with local customs or practices" in Benin or for "untitled property" in Madagascar.
- Set in place a land information system and a mechanism for managing land transactions at the communal level. The communes are charged with issuing certificates, recording transactions, and keeping registries up to date.
- Organize a link between lands covered by land certificates and the registration process. Even if land certificates meet the land security requirements of the great majority of rural stakeholders, there still should be bridges for obtaining a property title. Mapping resources should be shared with the state property department.

These policies represent a radical break with the earlier situation. They are still at the experimental stage.

---

products, provide income to a growing population, and exploit ecosystems in a sustainable way.

Farmers worldwide have amply demonstrated their adaptive capacities. Contrary to the Malthusian view, there are many examples that show that farmers know how to respond to land pressure by boosting the productivity of the land, sometimes at the cost of a decline in labor productivity and thus in standard of living, and sometimes through true agricultural revolutions that restructure the agrarian systems and increase the productivity of lands dramatically. But they can do so only within a favorable economic and institutional context and a land policy framework that is both appropriate and driven by the principle of fairness.

Incentive Agricultural Policies

There can be no agricultural development if the prices for farm products are too low for farmers to pay for their work while investing in productivity gains and ecosystem maintenance. The liberalization of trade in agricultural products creates unequal competition between farming systems with very different productivity and has the effect of marginalizing agriculture in the most difficult areas.

Incentive agricultural policies that provide farmers with a favorable economic environment, facilitate their access to means of production, and contribute to subsector rationalization are thus indispensable.

The Necessity of an Improvement in Farm Structure

There has been very little discussion about improving the structure of family farms in Sub-Saharan Africa, that is, not only the way in which the land is distributed between farmers and pastoralists, but also how agricultural plots and pasturelands are distributed in space. The land tenure debate has not been couched in these terms until now, no doubt because it is more popular to talk about poverty and inequality and small farmers. Another likely reason is the fact that the dominant farming systems, because they lack motorization, make the problem less acute. Nevertheless, land pressure, the emergence of landless farmers, competition with urban dwellers for land ownership, and the need for gains in agricultural productivity all call for serious consideration of the land structure issue. There are two main aspects to this issue: the size of family farms and what the parcel plan covers, and the spaces over which the family holds rights.

In areas where the space is saturated, the interplay of inheritances results in fragmented farms; in such cases, the land structure becomes increasingly unsuited to a vision of the family farm that emphasizes modernization and risk minimization. Furthermore, the conditions are rarely favorable for heads of farms to be able to easily revise their land structure by buying, leasing, or exchanging land.

Based on region and crop, what farm structures are viable for enabling farmers to derive a decent income and invest in farm modernization? What type of access should women and youth have to farming rights and property? What actions are needed to help young people who lack land, or who do not have enough land to make a living? How can alternatives to farming (food processing, agricultural services, and so forth) be promoted in rural areas or elsewhere, to give young people an opportunity to make a living and reduce the pressure on the land?

With respect to farm size, the dominant tendency among observers who are not agricultural specialists is to trust that rural exodus will solve the problem, but often it turns out to be insufficient and not necessarily equitable. Because of the magnitude of population growth, departures toward the city are not enough to relax the pressure on the land. Moreover, in countries where public and private investment is incapable of keeping pace with urban expansion, and where job prospects outside of agriculture are limited, the rural exodus exacerbates urban tensions. Last, for lack of a reliable alternative, those who leave the rural areas often want to retain access to land back home.

The challenge is to come up with land regulation policies that will not only lead to a better distribution of land based on needs, but also encourage the dynamics of production in a context where the necessary gains in productivity require viable farms while land pressure is causing their fragmentation. Isn't it time to begin thinking about farm structure policies appropriate to African contexts?[4] In the future, will there be—similar to the experience of rural communities of the Senegal River Valley or the Office du Niger—land commissions jointly managed by farmers to tend to the assignment of available lands or regulate land transactions in favor of family farms?

The problems raised by the distribution of farm parcels in the village, their size, and their groupings are, without exception, overlooked by land policies. And yet these problems are arising more and more frequently and in multiple forms. Thus, as part of their risk reduction strategies, farmers may prefer to have parcels located in different ecological settings (for example, dryer or wetter) in order to minimize their risks. Those who proceed with motorization may wonder about the size of their parcels; those who want to organize their work optimally may prefer that the parcels be grouped around their home.

Develop a New Conception of "Common Resources" for Ecosystem Management

Sustainable management of ecosystems requires regulating the exploitation of renewable natural resources; sustainability means that the level of exploitation remains below the annual productivity. Private property in no sense guarantees such regulation; there are many examples of this from the mining industry, resulting from either the quest for immediate profits or the lack of alternatives for poverty reduction. In addition, some natural resources do not fall within the province of private property, either as a result of their nature (resources that are dispersed, scarce, or irregular) or by choice of the society to define the resource as "common." The latter often applies in the case of forest reserves, pasturelands, wildlife, water points and their fish, and irrigation water.

Contrary to references to "the tragedy of the commons," we now know that sustainable management of common resources is possible, provided that institutions take on a role of responsibility. In many regions of the world, where local societies have historically established rules for regulating strategic resources that they have chosen to consider common, they continue to manage these resources in this manner. However, such "traditional" systems are often undermined by government interventions that fail to understand them or that reject them, by the arrival of new social groups who contest the rules, and by economic pressure. Furthermore, some resources that were once abundant and that were never regulated now need protection. In some cases, sustainable management requires strengthening or overhauling the ways of managing a common resource or even the creation of a new common resource.

One of the greatest difficulties is how to achieve harmony between ecosystem management (which involves not only farmers) and the management of farming systems by farmers themselves. Farmers are concerned about their farms, and about the ecosystems as well, but not in the same way. In the absence of gains in productivity or price increases that will enable them to live off the land that they cultivate, farmers are induced to clear new land, cultivate marginal areas prone to erosion, and overexploit natural resources, thus reducing available forage resources for pastoralism and the supply of available wood. The situation clearly depends on population density, the characteristics of the ecosystems and production systems, and the economic environment.

Many experiments have been carried out with external support to promote better management of village lands, control desertification, and conserve soils and water, with greater or lesser success depending on the balance between individual and collective interests that the experiment was able to strike. One of the greatest difficulties is how to reconcile short-term economic imperatives facing farmers and long-term environmental requirements. Another challenge is how to expand the scale, based on successful local experiments that garnered strong external support. At the same time, in areas where they are of vital importance, relevant achievements can strongly mobilize farmers who are ready to put considerable effort into common approaches to managing their village lands, in addition to working their fields.

The Need for New Land Policies

Land policies establish the principles and methods for managing rights to land and to natural resources located on the land (Lavigne Delville 2006). Thus, they define

- rights that are legally recognized and the procedures for recognizing them
- obligations or restrictions on how land and natural resources are to be used and ways in which land rights can be transferred
- the entities responsible for implementing land management, arbitrating disputes, and so forth.

A land policy may be set forth in a policy paper. It is then put into place through laws, decrees, and entities charged with implementation (at the national, regional, or even local level).

Currently, and all too often, the laws on land and resources remain out of sync, if not in outright conflict, with the day-to-day realities faced by farmers and pastoralists. Their rights to land and resources are not legally recognized and, in the event of a dispute, the arbitration process is not necessarily legitimate. This view is now widely shared, so there are many discussions aimed at updating land policy and reforming the law. A number of experiments are under way in various West African countries, designed to identify and formalize local land rights, formalize the rules negotiated for managing land and resources (local agreements), formalize land transactions (sales and leases), regulate contracts to provide inputs in the case of sharecropping activities or picking rights, negotiate local agreements on natural resource management, and develop communal land registries (box 6.2).

Efforts to come up with new land policies focus most often on the issue of land security, but also on conflict reduction, so that rural stakeholders can make optimal use of their space. To that end, it is important to clarify the rules and rights that meet with consensus at the local level and to provide them

BOX 6.2

Registration Policies: The Sole Solution?

Registration policies for land rights are among the standard recommendations in the area of land policy: local rights are informal and thus insecure, and as a result, producers are not motivated to invest. These rights, therefore, need to be formalized and property titles issued. Feeling secure and able to use the land as collateral, farmers will invest and achieve gains in productivity.

Many studies, both theoretical and empirical in nature, have disputed this seductive but simplistic argument (Platteau 1998a, 1998b): it is not true that informal rights are necessarily insecure. The fact of having a title is not sufficient for gaining access to credit; on family farms, the bulk of the investment is in labor, not capital, and property status is rarely the principal impediment to investment—price levels and the structure of industries often have a much more decisive impact. In addition, once "distortions" arise in other "markets" (for products, credit, or inputs), liberalization of the land market is highly likely to have negative effects on productivity and fairness. Finally, the task of developing a land information system is a costly operation that makes no sense unless consistent updating of the information can be assured, unless it is in the farmers' interest to have sales recorded, and unless the procedures are accessible, reliable, and inexpensive. Thus, a registration policy is not a universal solution.

Despite these widely noted findings, registration policies have taken on new vigor with the arguments advanced by De Soto (2005), who views formalization of land rights as a way to release the economic potential of the poor and generate an economic uptick. But while his analysis of the cumbersome nature of the procedures—and what this means for the great majority of the population in terms of excluding them from access to land rights—rings true, he does not address any of the questions raised above. Nor does he address the issue of how to take into account the diversity of local rights.

with legal recognition at the national level. An important part of the debate revolves around the following question: how far should one go in devising diverse regulatory responses, given the diversity of local situations to take into consideration?

On another note, the government needs to promote densification of rural spaces to better justify infrastructure costs, but without exacerbating poverty and the pressure on ecosystems, which raises the need to examine links between land policies and land use planning.

Farmer organizations are too often absent from the debate, sometimes because they have not been involved in the process and sometimes because they have not conducted internal discussions that would allow them to construct their own analysis and proposals and thereby fully participate. The involvement of farmer organizations is, however, essential so that they can advance the points of view and priorities of the rural population and to ensure that the choices made truly address their expectations and aspirations. In Senegal (CNCR 2004) and in Mali (AOPP 2004), farmer organizations have engaged in consultative processes and defined their positions. However, it bears mention that, to date, officials of farmer organizations have rarely addressed the issue of land inequalities between poor and rich farmers.

Another important issue concerns the financing of land policies, knowing full well that this is a sensitive matter in which the authorities do not necessarily want to involve external donors.

Given the challenges that exist, there is, to some extent, a need to invent solutions; appropriate responses can emerge only when the debate focuses on concrete problems, experimentation with possible solutions by stakeholders, and a search for the best compromises. Then, at the following stage, the challenge is to make a commitment to sustained action, which in itself is not an easy task.

Conclusion

In Sub-Saharan Africa, rural land policies must take into account the diversity of situations, contribute to balanced land use planning, provide land security for agricultural activities, address the periurban phenomenon, promote economic development and the reduction of inequalities, contribute to full consideration of environmental issues, address the issue of young people setting out to make a living, and more.

And they must do so in a specific context: a high population growth rate; difficulties in providing employment and income opportunities to new generations of rural youth; acute ecological problems at the local level (impoverished and fragile tropical soils) and also at the global level (greenhouse effect, disappearance of tropical forests) as a result of the clearing of "virgin" lands, which, moreover, are increasingly scarce in many countries; and a major challenge in the area of food supply, which requires spectacular improvements in farm productivity when available space is virtually exhausted.

In other words, the goal is to create—both in rural areas and in their interface with periurban areas—the right conditions not just for a green revolution (irrigation, improved seeds, fertilizer, pesticides), but for a double green one that will make it possible to achieve the necessary increases in land productivity, while simultaneously ensuring ecological sustainability by relying much more heavily on a keen understanding of individual ecosystems (Griffon 2006).

Above and beyond their contribution to meeting the challenges of food supply and land use planning, land policies fundamentally need to influence the choices made by society, in particular by establishing a social pact between government and the citizenry, defining the type of farms to be promoted and encouraged, and facilitating access to land and natural resources in ways that will lead to equitable economic development without creating a source of social or political conflict.

Yet the solutions to land-related problems lie not only in land practices and policies. A link to economic policies providing real support for family farms is also indispensable in order to ease the pressure on the land, absorb some of the population growth into agricultural and extra-agricultural activities, provide a future for the country's youth, and enable rural inhabitants to live with dignity on the land.

Notes

1. This chapter is greatly indebted to program documents of the French foreign aid agency's "Foncier et développement" program and the "Le foncier, un enjeu crucial aux multiples dimensions" report in *Grain de sel. 36* (http://www.inter-reseaux.org/); it restates certain parts of the introduction (Lavigne Delville 2006).
2. An analysis that focuses solely on private property and property markets cannot reflect the full diversity of land situations. A broader approach that encompasses both ownership and user rights is thus indispensable.
3. The urbanization ratio is 37 percent in Asia, 38 percent in Africa, and as high as 75 percent in Latin America.
4. In Europe, and particularly in France, farm structure policy has been one of the components of agricultural modernization policies. Using a range of tools, including the SAFERs (land planning and rural settlement organizations), farm structure policy has focused on promoting access to property for farmers, through land preemptions and preferential sale to young farmers at the start-up phase.

Bibliography

Amanor, K. S. 2005. "Jeunes, migrants et marchandisation de l'agriculture au Ghana." *Afrique contemporaine* 214.

AOPP. 2004. *La question foncière au Mali: Propositions paysannes pour une gestion pacifique et durable des ressources foncières au Mali.* Bamako.

CNCR. 2004. *Actes du séminaire national des ruraux sur la réforme foncière.* Dakar.

Compagnon, D., and F. Constantin, eds. 2000. *Administrer l'environnement en Afrique.* Karthala/IFRA, Paris.

De Soto, H. 2005. *Le mystère du capital: pourquoi le capitalisme triomphe en Occident et échoue partout ailleurs.* Flammarion, Paris.

Gonné, B., and C. Seignobos. 2006. "Nord Cameroun: les tensions foncières s'exacerbent." *Grain de sel* 36.

Grain de sel. 2006. *Le foncier, un enjeu crucial aux multiples dimensions.* Inter-réseaux développement rural. Paris.

Griffon, M. 2006. *Nourrir la planète.* Odile Jacob, Paris.

Hesseling, G., and E. Le Roy. 1990. "Le droit et ses pratiques." *Politique africaine* 40: 2–11.

Konaté, G. 2006. "Burkina Faso: une insécurité foncière féminine." *Grain de sel* 36.

Lavigne Delville, Ph., ed. 1998. *Quelles politiques foncières en Afrique noire rurale? Réconcilier pratiques, légitimité et légalité.* Ministère de la Coopération/Karthala, Paris.

———. 2006. "Quels enjeux pour les politiques foncières? Sécurité foncière, marchés et citoyenneté." *Grain de sel* 36.

Le Bris, E., E. Le Roy, and P. Mathieu, eds. 1991. *L'appropriation de la terre en Afrique noire: Manuel d'analyse et de gestion foncières.* Karthala, Paris.

Le Roy, E. 1997. "La sécurité foncière dans un contexte africain de marchandisation imparfaite de la terre." In *Terre, terroir, territoire: les tensions foncières,* eds. Blanc-Pamard and Cambrézy: 455–72. Collection Dynamique des systèmes agraires. Paris: Orstom.

Merlet, M. 2002. *Politiques foncières et réformes agraires: Cahier de propositions.* Réseau agricultures paysannes et modernisation/Fondation pour le Progrès de l'Homme. IRAM, Paris.

Platteau J.-Ph. 1998a. "Une analyse des théories évolutionnistes des droits sur la terre." In *Quelles politiques foncières pour l'Afrique rurale?* ed. Lavigne Delville.

———. 1998b. "Droits fonciers, enregistrement des terres et accès au crédit." In *Quelles politiques foncières pour l'Afrique rurale?* ed. Lavigne Delville.

Chapter 7

Innovation Systems and Support Mechanisms

Jean-Pascal Pichot and Guy Faure

African agriculture is innovative by nature, as shown by its capacity to integrate many species foreign to the continent as a complement to traditional crops and for the purpose of diversification. Projects and mechanisms for "modernization" have attempted to amplify the rate of change, initially based on a technically oriented, top-down vision of innovation. But consideration of factors in the socioeconomic environment has led to a rethinking of the position of African farmers, whatever their level of production, on the premise that careful assessment should permit a new approach to extension services and lead to innovations that are not just technical in nature but also economic and organizational, and that are appropriate to the constraints at hand.

Until now agriculture in Sub-Saharan Africa has managed to feed rural and urban populations in good years and bad, except in countries struck by political upheaval or serious climatic constraints. While the predictions of those who are pessimistic about Africa have not come true so far, a number of ecological, economic, and sociopolitical developments are heightening sources of anxiety about the future. The expectations of African societies in regard to African agriculture remain strong, so far as food security is concerned, but these expectations are growing increasingly complex in relation to food sovereignty, nutritional quality, rural employment, poverty reduction, sustainable development, and the management of renewable resources.

Can African agriculture meet these challenges and, if so, what will it take? Can one count on the endogenous, innovative capacities of family farms and the sociotechnical networks of the rural world? If not, can exogenous and exotic technologies be promoted (such as plant and animal GMOs—genetically modified organisms—or new cropping patterns, for instance) to expand production, or can agribusinesses that are well supplied with financial and human capital be relied on to reach the level of agricultural competitiveness achieved in emerging countries? Is it not necessary to accept interactions between these different dynamics in order to construct a range of responses, based on the local political, economic, and ecological situation and the capacity of stakeholders in the agricultural sector to adapt to demand and opportunities (Courade and Devèze 2006)?

This chapter seeks to answer these questions based on a review of innovation processes in Africa, primarily with examples from francophone Africa, in light of the authors' professional experience.

Great Diversity, Shared Challenges, a Changing Context

Agriculture in Sub-Saharan Africa is, in fact, characterized by a great diversity of situations given the natural settings (climate, soils, biodiversity, and so forth) in which farming takes place, the people involved in developing agriculture, and historical background. This diversity of local know-how, "land-capital-land" combinations, and performance is often underestimated, not only by politicians and those involved in official development assistance, but also by researchers and nongovernmental organizations (NGOs) (Pichot 1996).

All Sub-Saharan agriculture faces considerable challenges: climate change, evolution of agroecological systems, the increasing expense of fossil fuels and imported inputs, population growth and urbanization, and the evolution of markets and government policies.

At the same time, substantial transformations—and thus uncertainties—are affecting international trade: witness the growing role of trade between countries of the South polarized by the major emerging powers, as well as price tensions resulting from the processing of biofuels, such as maize in the United States, sugar in Brazil, and palm and other oils in Indonesia. The old bipolar relations with European countries built on commercial channels, official development assistance, and political and cultural ties are profoundly affected by these developments.

Finally, agriculture in Africa, as elsewhere, faces new quality requirements from consumers: health safety (absence of pesticide residues in vegetables, absence of parasites in fruit, absence of microbial toxins in cocoa and coffee), nutritional quality (Delpeuch 2007), environmental quality (sustainable production and processing systems, eco-certifications), and even social factors (child labor, fairness, democracy). These requirements do not uniformly apply to all products, but the growing power of quality standards can be observed on the international markets and also on the urban markets of certain countries of the South, including some African countries, where large-scale distribution is growing more common.

African agriculture—long known for its capacity to produce useful raw materials for European industries (cotton, coffee, cocoa, wood), and also known today for its ability to feed the cities of the South (Bricas 2006) and maintain rural employment—does not, however, enjoy a very favorable brand image. Often viewed as high-spending and little disposed toward modernization, African agriculture has witnessed a decline in the assistance and support services that it used to receive, as a result of the adjustment plans imposed on African countries by the International Monetary Fund and the World Bank (Felix 2006).

African Agriculture Caught in Transition between Endogenous Dynamics and Exogenous Support

Without attempting to do the work of historians, today's agronomists and specialists in animal husbandry are well aware that cassava, potatoes, maize, tomatoes, mangoes, citrus, horses, and humpless cattle are not indigenous species and that their adoption and dissemination have been going on for at least two centuries, often with no government intervention. The dissemination of these "exotic" species diversifies the farmers' production systems but without causing the disappearance of local species of grain (millet, sorghum, rice), root crops (yams), perennials (oil palm, safou tree, cola tree), or livestock, for which farmers see to domestication, breeding, and biodiversity (see colloquiums on the biodiversity of sorghum held in Bamako in 2007 and the domestication of yams held in Montpellier in 1992).

Unfortunately, the decline of agricultural statistics makes it impossible to develop a precise image, apart from ad hoc surveys, of the recent evolution of these products, which now more than ever fall within the informal economy.

The introduction and development in the past century of crops that are part of a trade-driven economy—such as groundnuts and cotton in the savanna regions, or coffee, cocoa, and rubber trees in the forest—have profoundly modified Sub-Saharan agriculture. Instituted before the creation of independent nations and sometimes under duress, the changes linked to these crops have continued up to the present day. They are easier to track, since they appear in the data collected in the process of monitoring industrial activities and international trade. Strongly supported by newly independent governments, these crops have given rise to integrated industries, providing stakeholders with a secure framework conducive to technical, economic, and social innovation. These integrated industries deserve credit for family farms' mastery of mechanized production techniques relying on external inputs, as well as for the creation of trade organizations and for income generation in rural areas.

In the current conditions, these industries remain the foundation, along with wood, of agricultural exports from Sub-Saharan Africa and continue to form the income base for many farm families. However, this historical model of industries exporting slightly processed (or unprocessed) agricultural products now appears to be a source of anxiety over income and employment.

The volatility of the international markets, combined with the incapacity of African governments to protect their agricultural sector or to finance agricultural support services and rural credit, has led through various channels to decapitalization of rural family farms. To various degrees, this decapitalization has affected:

- social and human capital, caused by the breakup of large families working together on farms possessing an ample supply of inputs; the migration of young people of working age toward cities and foreign countries; and the weakness of agricultural education, research, and support services
- biological and ecological capital, stemming from declining pastoral resources and growing desertification in the North, disappearing forests, aging orchards and plantations, and difficulties in replenishing them further south
- operating capital, attributable to aging agricultural equipment or the sale or nonreplacement (for lack of appropriate credit) of same, and the obsolescence of some agribusinesses (oil mills, for example)
- land capital, resulting from fragmentation of farmlands in densely populated areas, limiting their capacity to accumulate or even survive

However, in reaction to the emergence of large cities and urban markets, new forms of production, marketing, and processing have developed over the past 20 years, sometimes at a considerable distance from the cities. Fruit trees, market gardening, floriculture, short-cycle stockraising (poultry, swine, aquaculture), and milk production all supply markets with fresh products and assorted processed products in response to the expectations of urban consumers. This periurban and sometimes intraurban farming activity innovates constantly, most often without much support from government services or health controls over its products.

Thus, in rural areas located far from urban centers, just as in periurban areas, family farms today all focus on their relationship to markets (the product market, the capital market, the labor market, and, more and more frequently, the land market). The differences have to do with the methods of access to these markets and to natural resources, through social and trade networks that facilitate the flow of information and products in various ways, and through agricultural support services that facilitate the task of adapting systems of activities in line with changes in the context.

How can we best take into account the full diversity of this agriculture, which should for once and all cease being termed "traditional," to build appropriate development policies?

To help family farms innovate and adapt to an uncertain future, it is, of course, necessary that the arrangements governing access to land, water, and pasturage be stabilized and made secure. Access to credit is also necessary so that family farms can obtain the production, transport, storage, and processing equipment needed to improve their productivity. But one thing may be needed above all else: political, economic, organizational, and scientific officials must reach agreement on what innovation in agriculture actually means, what the mechanisms of innovation are, and what support is possible in the current context of "less government."

What Does Innovation in Agriculture Really Mean?

The innovation processes in African agriculture—endogenous, exogenous, or mixed—have not given rise to many analytic works, or to capitalization. Part of this stems of course from weak human resources in the economic and social sciences at the research and teaching institutions involved but such research also requires a multidisciplinary approach and close proximity to a society in movement, neither of which is necessarily sought out by researchers and those who evaluate or finance the research. To analyze the processes of innovation, one needs to take the risk of placing oneself at the heart of the change, that is, at the interface between innovator and society, while at the same time maintaining theoretical references and a strong capacity to reflect on the generic applicability of changes and their consequences.

Early on, Schumpeter (1935) defined innovation as a new combination of inputs, which could be reflected in a new product, a new way of producing, the creation of new outlets, or access to new resources. Consistent with this definition, the traditional view is that innovation is radically different from invention and that it is varied by nature: technical, economic, organizational, social, and, moreover, generally composite. Thus, technical innovation typically goes hand in hand with an organizational innovation. It could even be said that the former is embedded in the latter.

Innovations in agriculture may of course focus on production processes; after all, agricultural research is often expected to cover technical issues only (Pichot, Sedogo, and Deguine 2006). However, upstream control of access to inputs (land, water, labor, credit, seeds) and downstream control of postharvest product care (drying, fermentation, sterilization, storage), or even of product processing or market placement, may turn out to be much more important than production itself when the goal is to improve the productivity and competitiveness of the agricultural sector and reward family labor (box 7.1).

Innovation may come as an add-on to longstanding systems or it may represent a break with earlier ways. Its roots may be exogenous, brought by the

BOX 7.1

Example

One example of this family labor is the project to promote industrial production of cassava in Toumodi, Côte d'Ivoire, in the 1970s. This project was designed to supply the markets of Abidjan. Once the technical success of motorized production of cassava was assured (to which agricultural research had contributed), the project nevertheless ran into the social realities of women's trade networks supplying urban markets with processed products that meet consumers' expectations.

world of technicians, or endogenous and brought forward by farmers. Often, though, a combination of the two comes into play, either through the collective efforts of farmers and technicians (in which case the innovation is collaborative in nature) or through delayed implementation (an exogenous innovation proposed by a project at a given time, but rejected by farmers, may be resurrected and transformed by an individual or a group many years after the project has ended), or yet again through implementation in different locations (an endogenous innovation in one area may be taken up by technicians and disseminated in another area where it is unknown). An innovation may emerge in very diverse contexts and make sense only when one analyzes the social and political context in which it emerges. It may develop, and may then lose relevance if the context evolves. Innovations that are part and parcel of the strategies of certain stakeholders to achieve their objectives or strengthen their position raise issues of power and negotiation within society. To illustrate the complexity of the phenomenon, Olivier de Sardan (1993) defines it as "a totally new graft between two floating systems, in one place, via the ferryman."

In the case of industrial crops, a change of variety (cotton, sugar cane, maize, soy) can be implemented without difficulty since the seed production and distribution chain already exists and because the stakeholders affected by the change are clearly identified within the sector. In the case of food crops such as rice, a change of variety will raise questions upstream concerning production and distribution of the seed. But downstream of production this innovation will also raise questions about the technological properties of paddy rice in the processing chains (artisanal or industrial), and about the response to the domestic consumer's expectations (cooking and organoleptic qualities). Thus, the success of the innovation also depends on returns and other advantages and risks as perceived by other sector stakeholders (box 7.2).

Mendras and Forse (1983) propose five factors for assessing the adoptability of innovations, as follows: the relative advantage provided by the innovation compared with the initial situation; its compatibility with the system in place;

BOX 7.2

Example

In Upper Guinea, as in Burkina Faso, the artisanal hulling and steaming of rice by women represents a major element of the attractiveness and competitiveness of locally grown rice in comparison with imported rice. Simple technologies—based, for example, in the case of steaming, on the artisanal recycling of metal cans—provide a way to improve the productivity of their work. This technological innovation makes the greatest sense in the case of the Mogtedo cooperative in Burkina Faso, where it fits into an entire system, controlled by stakeholders, for regulating the volumes and prices of steamed rice placed on the market locally, some 90 kilometers from Ouagadougou.

its greater or lesser complexity; the ease of trying it out in the stakeholder context; and the possibility of observing its use by others. These factors incorporate the degree of complexity and the risk level for producers. Darre (1996) attaches great importance to exchanges and discussions within producer groups (sociotechnical networks) in explaining the dynamics of innovation. Olivier de Sardan (1998) stresses the role of social agents who occupy a more or less recognized position in local society and through whom the innovation passes. Furthermore, an innovation that is adopted produces indirect and generally deferred effects on the local social structure by serving certain interests and running against others; it can thus consolidate the social structure in place or, conversely, promote a new deal of the cards.

In addition, the diversity of climatic, ecological, geographic, social, and political conditions can limit the scope of validity of an innovation: North-South technology transfers, and now, East-West transfers (for example, between Brazil and Africa) may, of course, feed information into the innovation processes but are not a full substitute for them, contrary to what was generally believed 30 years ago about appropriate technologies.

Chauveau, Cormier-Salem, and Mollard (1999) point out that supply and demand for innovation revolve around interactions between stakeholders over technical matters; that innovations take root through composite networks reflecting the heterogeneity of socioeconomic units; and that the relationship between innovation and the economic, social, and political environment shows that the processes are not linear. Thus, stakeholder knowledge—both ancient and new—and stakeholder actions based on strategies that are not always explicit serve to modify the relationships; influence society's rules of operation; and generate innovation processes involving accumulation, adaptation, and rearrangement of various inputs.

The complex interrelationships among stakeholders form the basis for proposing the concept of an "innovation system," which can be "defined as a network of organizations, enterprises, and individuals focused on bringing new products, new processes, and new forms of organization into economic use, together with the institutions and policies that affect their behavior and performance" (World Bank 2006) (box 7.3).

In the agricultural sector (in the broad sense), these innovation systems cannot bear fruit (in terms of productivity, competitiveness, employment, and so forth) unless the changes they support are consistent with the values and rules underpinning rural society and unless they embrace the totality of stakeholders. Unlike other sectors where protecting information is a key to success for any innovating enterprise, in the agricultural sector the dissemination of information and knowledge is necessary for ensuring social consolidation and the sustainability of innovations. Agricultural support services and trade organizations—not to mention the new information and communication systems (rural radio, specialized press, electronic forums)—can play an important role by providing stakeholders with meaningful information on markets, opportunities for financing, and management (box 7.4).

Factors of Change and Modes of Innovation

Many factors seem likely to push technical or social practices to the point of change (Lélé 1989).

BOX 7.3

Example

No one should overlook the efforts made in the past 20 years by international agricultural research, as well as by NGOs backed by various European aid agencies, to promote corridor cropping between rows of *Leucaena* sp. and cover plants (*Mucuna*, for example) in the Bight of Benin as a way to improve soil fertility and fight weed infestations. These efforts have often sparked the interest of farmers, but the absence of immediate economic repercussions, combined with a workload increase, has impeded permanent adoption of these technical systems despite their being considered highly effective by researchers (Carsky et al. 2003).

Systems of direct seeding and cover crops follow a similar, deliberate intellectual path at the outset; their dissemination in savanna areas has been supported by some cotton companies (in Burkina Faso, Cameroon, and Mali) and donors, based on the normative top-down model. Only the future can say whether these technical recommendations originating in Brazil will meet the expectations of African cotton producers.

BOX 7.4

Example

In Indonesia, mobile telephony allows rural producers of cocoa to closely monitor price trends and thereby more optimally negotiate the farm gate price of their product with buyers. As a result, and because of competition between buyers, these producers are able to pocket as much as 75 percent of the world price, whereas African producers receive less than 50 percent.

The pressure on resources, and especially on land, particularly in isolated rural areas with low rainfall, leads to ecological decapitalization (overexploitation of ligneous resources and decline of biodiversity, soil fatigue and erosion) that, to observers, often raises the fear of an irreversible crisis of agroecosystems. Oftentimes, however, changes in practices go in the direction of intensification of production based on an increase in labor per unit of surface area: examples include early seeding and careful manual weeding of grains in the Senegalese Groundnut Basin; plantings of trees (for example, mango and cashew in Benin, Côte d'Ivoire, Mali, and elsewhere); harvesting and storage of cereal straw for livestock fattening operations in Mali; agroforests in Guinea; bocage landscape in the Bamiléké region; and flood recession crops in the flooded areas of North Cameroon (box 7.5).

Such innovations relying on family labor do not require financial risk-taking in the short term, but the social cohesion of family groups may run into issues of hard work and payment of family workers. In addition, the medium- and long-term ecological sustainability of these labor-intensive systems is not always apparent, even when they integrate mixed cropping-stockraising or agroforestry systems. These systems, based on a transfer of fertility within the village space generated by herd movements and, earlier, by fallow, reach their limits when land saturation causes agropastoralists to send their herds on long transhumant itineraries for lack of forage plots, fallow land for grazing, or common areas used for pastoral purposes within the reach of villages.

In extreme cases where agricultural production systems no longer lend themselves to the reproduction or cohesion of family groups, the success of family labor may lie in paid, temporary agricultural activities; in nonagricultural activities; or in prolonged migrations toward production hubs (Office du Niger, cotton-growing basin of Mali, Senegal River Valley, Lake Alaotra in Madagascar, coffee- and cocoa-producing forest areas), cities, or foreign countries. The ability of family groups to effectively use the income derived from emigration is, moreover, one of the challenges to the social sustainability of these expanded systems of activities.

BOX 7.5

Example

In the densely populated Korhogo area of Côte d'Ivoire (Dugué 2001), in Mali at the heart of the cotton-growing basin (Dufumier and Bainville 2006), and in the Machakos district of Kenya (English, Tiffen, and Mortimore 1994), farmers have come up with endogenous solutions for agricultural problems for many decades. Some projects have been able to support these dynamics by providing training to farmers and encouraging innovation processes that build on known techniques (planting *Acacia albida* in North Cameroon, installing stony cordons in Kita in Mali). These innovative practices for managing an ecosystem's fertility (relying on ligneous plantings or mixed cropping-stockraising systems) perfectly illustrate the divide between the reality in the field and the conventional comments on the inevitable degradation of agricultural ecosystems from the effects of population growth (Jouve 2006).

The lure of local, national, and regional markets is a powerful engine of change that may affect actual practices. For optimal functioning, this requires, however, that transportation infrastructure (roads, bridges, ferries, railroads, airports) and storage facilities (warehouses, silos, refrigerated spaces) be in "good" condition, that "undue" charges remain reasonable, and that transport enterprises enable commercial sectors to develop sustainably both upstream and downstream. The question of "territorial continuity" remains an issue in many countries during the rainy season (Magrin 2001). In addition, the rising cost of transportation (as fuel prices climb) could result, if it continues, in a contraction of the geographical areas supplying cities, processing plants, and ports of shipment (box 7.6).

Innovations sparked by the lure of the marketplace often involve recourse to inputs (seeds, fertilizers, pesticides) and mechanization under certain conditions: prices offered on national or international markets that are high enough to reward producers; national and regional agricultural policies that are conducive to the creation of input distribution and rural credit networks; governments that are prepared to implement quality controls for such inputs; and road systems that are in good condition, particularly in areas situated far from cities. But even in areas where these favorable conditions exist, the distribution of underdosed fertilizers, ineffective pesticides, and even pesticides that are toxic to users is not, unfortunately, a rare occurrence. In this type of agriculture, especially when it is structured by the proximity of urban markets, the new communication technologies facilitating a real-time flow of information enable producers—and, more often, intermediaries—to monitor prices

BOX 7.6

Example

In Côte d'Ivoire in the 1980s, following a political decision to ban imports of maize, the production of grain-maize spread rapidly in cocoa-producing areas as a way to supply poultry-raising operations in Abidjan with poultry feed, which had been lacking. This response to the urgent needs of poultry farmers was supported effectively by cocoa-marketing networks that benefited from the proximity of the city, the existence of infrastructure, and the necessary contacts.

In Chad over the past decade, swine raising has undergone substantial growth in order to supply the urban markets of southern Cameroon with brochette meat and ease the shortage caused by a swine fever epidemic that decimated stockraising operations.

In southern Mali in the 1990s, potato farming (a crop requiring irrigation and expensive seed stock that is difficult to import) was unexpectedly developed by men as an off-season crop, in competition with the bottomlands rice grown by women. During the same period, Office du Niger rice growers in Mali demonstrated the ability to diversify their production systems (stockraising, off-season market gardening) and obtain fertilizer in Côte d'Ivoire at a time when sales of inputs by government agencies had ended.

---

and quantities of perishable products (vegetables, fruits, fish) available for sale, and thus more effectively seize market opportunities.

For farmers, technical changes that rely on expensive inputs mean taking a financial risk and accepting a dependency on supply channels that cannot be individually controlled. Any decline in the use of intensive production techniques, as witnessed today in most sectors, greatly depends on the stakeholders' perception of economic contingencies: the instability of local and international markets, unexpected imports of low-priced competing products (Asian broken rice, European potatoes and onions, cheap European chicken parts, Brazilian maize), uncertainties about the distribution of inputs in rural areas, and so forth. The efforts made by farmer organizations to exert influence on international agreements of the WTO, bilateral agreements, and partnership agreements between the European Union and the African, Caribbean, and Pacific Group of States, as well as on national and regional agricultural support policies and agricultural supply companies, seem justified because the stakes are so high.

The proximity of urban markets leads to the emergence, around and within major population centers, of original forms of agricultural production, described as periurban farming. These forms of agricultural production (orchards, short-cycle stockraising, processing workshops, fish farming) are highly reactive to consumer expectations, they are located close to input distribution companies, they generate employment, and they are apt to spark the interest of urban investors. Periurban farming may usefully recycle a portion of urban waste, but it may also involve less than perfect mastery of pesticide use, thereby creating public health problems, or have a very negative impact on groundwater quality (as in the case of the Cap Vert peninsula in Senegal, for example), which would undermine its sustainability.

The Impact of Large Public or Private Development Operators and Large or Midsize Distribution Operators

In Sub-Saharan Africa, government intervention on behalf of many integrated sectors (encompassing production, processing, and marketing) has led to the creation of numerous specialized development companies, differentiated by product (cotton, coffee, cocoa, rice, sugar), to which governments have often assigned a mission of public interest involving infrastructure, rural credit, input purchasing and distribution, information and training for producers, support to trade organizations, support to innovation processes in conjunction with completed research, or extension.

These export sectors have significantly contributed, until recent years, to the development of rural employment and the dissemination of animal traction, mechanization, herbicides (improving labor productivity), chemical fertilizers (improving the productivity of the land), and pesticides. The innovations brought by these integrated sectors have irrigated all the production systems located in the producing areas, improved the overall productivity of the agricultural sector, and greatly contributed to the emergence of trade organizations.

Despite the results obtained, these development companies have lost, sometimes all at once (Lake Alaotra Development Company in Madagascar) and sometimes gradually (Ivorian Textile Development Company in Côte d'Ivoire), all or some of their capacity to intervene in economic matters under the structural adjustment and agribusiness privatization plans. This has sometimes released the capacity of the private sector or trade organizations to advance an initiative, especially for the distribution of inputs or short-term credit (Zoundi, Hussein, and Hitimana 2006), but overall, there is no longer any guarantee in many cases that a development company's mission of public interest will be fulfilled, notwithstanding its importance to the innovation process: research, advisory and support services, farm equipment loans, training for rural populations (Wampfler 2006), and road and bridge maintenance in rural areas.

The very serious financial difficulties facing public institutions engaged in research, teaching, or support for agriculture make it difficult to question programs that often continue to favor the acquisition of generic knowledge and the dissemination of standard technologies to the detriment of tailored approaches that reflect local contexts. Renewed attention to farmer knowledge and know-how and farmer-based projects, and their crossbreeding with knowledge produced by technicians and adapted to the new expectations of operators (producers, processors, traders), are, nevertheless, necessary to promote the spread of innovations anchored in local social realities that also take market constraints into consideration.

The private sector of large enterprises can also be a source of innovation. However, the presence of companies specializing in large or midsize distribution in African cities is already generating new problems in some countries (South Africa, Kenya), because these operators impose standards on their suppliers with which small producers on family farms are not familiar. The small producers' difficulties in meeting such standards may disqualify them and exclude them from these modern channels for supplying cities, to the benefit of a limited number of "modern" producers that are capable of satisfying the quantity, quality, and delivery time requirements, and who are also able to make the necessary investments to comply with health standards.

These new requirements of large-scale distribution also affect export industries dealing in fresh products (vegetables, fruit, fishery products). They have produced drastic changes in the supply structure in many parts of Africa. For example, in the green belt of Dakar, small market gardeners are growing less and less for export, and they are increasingly becoming the employees of large, capital-intensive farms. The formative, economic, and social effects of these new operators still appear to be underestimated by trade organizations that are ill-informed about these changes and these new quality requirements.

Evolution of Support Mechanisms for Innovation Processes

While innovations are often exogenous to project and government agency interventions, this in no sense diminishes the fact that extension and research have also contributed to rural development. The evolution of concepts and institutional mechanisms of support for innovation in rural Africa over the past 40 years provides many useful lessons for developing proposals for the future.

In the 1960s, "agricultural development" was based on a linear, or top-down, transfer of technologies. Research generated techniques, extension disseminated the techniques through information and training, agroprocessing increased the security of operations upstream and downstream of production, and producers adopted the new approaches and modified their practices to boost their output and their income, or even for the sake of "modernizing" their farms.

This linear process of innovation was sufficiently effective to induce major modifications in production systems in certain situations where ecological conditions were favorable or agricultural policies created a context conducive to the development of certain industries (cotton, most symbolically, in the savanna regions, and palm oil, coffee, and cocoa in the forest regions).

But in many less favored areas (with more difficult agroecological conditions, higher economic risks, or a lack of well-structured industries), this intensification model of the green revolution was a failure, because the proposals to intensify the output of partially marketed products were out of sync with the real-life situations faced by farmers, thus posing a substantial challenge for research and development operators.

Expanding on this approach, until the late 1990s the World Bank promoted the Training and Visit method for organizing the work of government agencies involved in agricultural extension in developing countries. Based on the idea of providing information to farmers and demonstrating in test plots the standardized production techniques developed by research institutions, these top-down interventions did not attempt to address the diversity and environment of farms. They focused mainly on streamlining the way in which the work of government extension agencies is organized, without, however, achieving their objectives in this area.

However, in the 1970s and 1980s, other approaches gradually emerged, based on the concept of Research and Development (R&D), which Jouve and Mercoiret (1987) define as "real-scale testing, in close cooperation with farmers, of technical, economic, and social improvements to their production systems and ways of using their environment." These R&D approaches replaced the linear scheme (research-extension-producer) with a triangular relationship among the three types of actors at all stages of the innovation process, and they expanded the scope to encompass the producers' environment. They recognized that full development of an innovation must indeed take into account the real conditions in which agricultural production occurs, and that its adoption will depend on producers' objectives and means.

These approaches thus logically attached substantial importance to systemic critical analysis of production systems and practices at different levels of organization, to identify obstacles and the margins of progress likely to stimulate on-farm trials (and, if necessary, research further upstream). On-farm agricultural trials resulted in an abundant literature on the conditions and methods of such trials, as well as on their usefulness and their limitations (Triomphe 1989). Trials focused most often on technical improvements at the farm level, with a varying degree of participation by producers in programming the activities and assessing the results. The processes of adaptation and collective mastery of the innovation by producers (Lefort 1988), as well as the dissemination of results beyond the target population of R&D projects, failed to yield a significant scientific output concerning research and development approaches.

In the late 1980s, heavy criticism was leveled at research and development, revolving around the length and cost of the appraisal phase, the weakness of proposed actions, the exploitation of producers during the trial phase, and the limited field of validity of the results obtained. This criticism led to more participatory approaches designed to strengthen the individual and collective learning process for producers and the interactions among producers, technicians, and researchers. Thus, action-research in rural areas came to involve the explicit construction of a partnership between research and other stakeholders aimed at solving a jointly identified problem. In anglophone areas, the methods also evolved. The farming systems research projects of the 1970s and 1980s (Norman and Collinson 1985) were succeeded by rapid methods of participatory research (Chambers, Pacey, and Thrupp 1989), which gradually shifted the emphasis to rapid rural appraisal, participatory rural appraisal, and, more recently (Lavigne-Delville, Sellamna, and Mathieu 2000), participatory learning action.

Also during the 1980s, government extension agencies were greatly affected by government divestiture of some of the support services targeting rural areas, as well as by the influence of wide-dissemination approaches, notwithstanding the evidence pointing to inappropriate transfers of standardized technologies and technical packages poorly suited to the needs exhibited by a diverse spectrum of farms (Rivera 2003). The needs are very different between a small farm struggling to achieve food self-sufficiency for the family unit and a larger, well-equipped farm that sends a large share of its output to market.

The identification and implementation of new methods of providing support to producers has thus sparked initiatives by different stakeholders in recent times. For example, the Neuchâtel Initiative, backed by various donors from the international community, has since 1995 analyzed past and current experiments in agricultural extension in order to draw relevant lessons.[1] But alternatives to the wide-dissemination model have been under discussion for much longer.

As early as the late 1970s, research and development programs in Senegal (Benoit-Cattin 1986) came up with methods of providing technical and economic advice that consisted of proposing a set of solutions tested jointly with producers, after having conducted with them an analysis of their situation. This method of support, dubbed "management advice" by its authors, takes into account "the farm's entire situation and seeks, in dialogue with farmers, a path to progress that is often spread over several years" (Kleene, Sanogo, and Vierstra 1989). Subsequently, other ways of providing advice were tested in tropical areas, some of them patterned on the French approaches to farm support embodied by the Centers for the Study of Agricultural Techniques and the Institute of Management and Rural Economics (Legile 1998). These approaches aim to build the capacities of producers and make them more autonomous and thus better able to manage their activities and promote innovation on their farms.

Then, in 2001, a workshop was held in Bohicon, in Benin, with the participation of operators, researchers, and farmers involved in providing advice to family farms; this workshop served as a forum for developing an overview of lessons learned in this area and identifying future challenges (Dugué and Faure 2001, 2003).

In the 1990s, in view of mounting evidence of the private sector's limited capacity to spontaneously take over the government's role in providing services to producers, special emphasis was placed on efforts to strengthen producer organizations. The latter have something of the nature of the private sector, with hybrid forms of organization and operation. They handle multiple functions (Bosc et al. 2003): economic functions related to the production, processing, and marketing of products or to natural resource management; and social functions in terms of defending the interests of members, sharing information, building capacities, and providing coordination between local and global actors.

Despite the many difficulties they run into, these organizations are gradually building the capacity to formulate specific demands, to negotiate locally and regionally with other actors, and to develop new activities and services on behalf of their members. They represent a major institutional innovation, and they participate in the development of future social and technical innovations (Mercoiret 2006). Despite greater attention to farmers' demands, however, it is not clear that the largest organizations and their umbrella agencies are better prepared than the government support services of old to address the farmers' needs in terms of innovation.

The persistence of the dominant, top-down models of innovation, the tendency to replicate the hierarchical structure of certain rural societies in farmer organizations, and the habit of recruiting officials who are poorly trained to work with producers (Opio-Odongo 2000) are some of the handicaps that organizations will need to learn to overcome, which can take place only over time. In addition, the harmful effects of support lavished by European or North American experts who are not well informed about local policies and social realities, along with misguided interventionism by some international organizations, pose difficulties that, unless they are addressed, may result in repetition of the errors of the past (standardized solutions, negation of diversity).

Conclusion

The current economic and institutional conditions of African agriculture call for a new look at the ways in which innovation systems are supported. The results obtained by researchers should no longer be viewed as the primary source of technological changes to be implemented by economic operators based on a linear, top-down model. Recognition and mobilization of localized farmer knowledge and know-how should take center stage and form the basis for an overhaul of agricultural research and support systems. All stakeholders are called upon to build new contractual relationships, public-private partnerships, or other mechanisms that fit local contexts and involve multiple actors as a way to characterize problems, identify and test solutions, develop tools for steering the process of change, and assess the results of this process.

In some cases, this may require that farmer organizations strike an agreement with industrial operators to develop new production systems that will improve the supply of products of good technological quality for factories and ensure that value added is shared equitably, without neglecting environmental externalities.

In other cases, this could require finalized research results, to be implemented with ad hoc organizations or local communities through consultative frameworks involving multiple actors, in order to resolve issues of shared and sustainable resource management (water, ligneous and herbaceous biomass, and the like) and conflicts between social groups (farmers and pastoralists, for example).

Or again, this could involve an NGO or a research-development-training center that provides support to a network of farms and agricultural enterprises as a way to promote new production methods apt to improve technical and economic performance.

This conception of innovation, rooted in partnership, does however raise a number of questions that have yet to be resolved.

First, the decline of government financing has led to a crisis of recruitment and operations in research, training, and advisory systems. To address this situation, other sources of financing need to be identified as expeditiously as possible through new institutional arrangements. Yet the contributions of the private sector remain limited: in some cases such contributions are difficult to mobilize because the sectors are not well structured (this holds true of most products destined for domestic markets); in other cases, they are unlikely because the sectors are in crisis (for example, the cotton and cocoa industries). Is it possible to forge relationships with foundations (Rockefeller, Bill and Melinda Gates, Farm), which now bring to the table greater funding than all international aid combined, in order to promote innovation systems?

Second, ownership of the results generated by such partnerships deserves attention inasmuch as the resulting knowledge and innovations may be deemed private property by some actors if they are considered critical to their competitiveness and therefore not subject to wider dissemination. There is an ethical imperative to ponder rules for governing these issues, tailored to specific situations.

Third, how can an innovation's social and environmental impact on rural societies and resources be measured so that stakeholders can control the effects of their actions, take stock of the results they obtain, and justify the investments made? The lack of a culture of external monitoring and evaluation persists among the actors in innovation systems, but there is also a lack of adequate information systems and financing mechanisms for these "watchdogs of change."

Fourth, skills development (for researchers and technicians, but also for farmers and the leaders of producer organizations) remains a challenge in creating balanced and effective partnerships. However, the mechanisms of agricultural education and skills development in rural areas and the systems for providing advice to farms are losing momentum. This situation requires, in tandem with support to agricultural trade organizations, new efforts to identify mechanisms of intervention that will benefit structures such as rural family schools, farmer universities, and systems of support and advice managed by producer organizations or private enterprises.

Note

1. http://www.neuchatelinitiative.net.

Bibliography

Benoit-Cattin, M., ed. 1986. *Les unités expérimentales du Sénégal.* ISRA, CIRAD, FAC, Montpellier.

Bosc, P. M., K. Eychenne, K. Hussein, B. Losch, M. R. Mercoiret, P. Rondot, and S. Mackintosh-Walker. 2003. "Le rôle des organisations paysannes et rurales (OPR) dans la stratégie de développement rural de la Banque mondiale." Stratégie de développement rural, Document de base 8. World Bank, Washington, DC.

Bricas, N. 2006. "Des marchés alimentaires urbains en plein développement." *Grain de sel* 34–35: 30–31.

Carsky, R., et al. 2003. "Amélioration de la gestion des sols par l'introduction de légumineuses dans les systèmes céréaliers des savanes africaines." *Cahiers de l'Agriculture* 12: 227–33.

Chambers, R., A. Pacey, and L. A. Thrupp. 1989. *Farmer First: Farmer Innovation and Agricultural Research.* Intermediate Technology Publication, London.

Chauveau, J. P., M. C. Cormier-Salem, and E. Mollard, eds. 1999. *L'innovation en agriculture, questions de méthodes et terrains d'observation.* IRD, Paris.

Courade, G., and J.-C. Devèze. 2006. "Des agricultures africaines face à de difficiles transitions." *Afrique contemporaine* 217 (1): 21–41.

Darre, J. P. 1996. *L'invention des pratiques dans l'agriculture.* Karthala, Paris.

Delpeuch, F. 2007. "Le système alimentaire mondial à un carrefour." *Cahiers de l'Agriculture* 16: 161–62.

Dufumier, M., and S. Bainville. 2006. "Le développement agricole du Sud-Mali face au désengagement de l'Etat." *Afrique contemporaine*, 217 (1): 121–33.

Dugué, P. 2001. "Dynamique de plantation et durabilité des systèmes de cultures pérennes en zone de savane de Côte d'Ivoire." Colloque sur l'avenir des cultures pérennes en zones tropicales humides, Yamoussoukro, November 5–9.

Dugue, P., and G. Faure, eds. 2001. *Le conseil aux exploitations familiales. Actes de l'atelier.* Bohicon, Benin. November 19–23.

———. 2003. *Le conseil aux exploitations familiales. Actes de l'atelier.* Bohicon, Benin. November. CD-ROM. CIRAD–IRAM–Inter Réseaux, Paris.

English, J., M. Tiffen, and M. Mortimore. 1994. *Land Resource Management in Machakos District, Kenya, 19301990.* Report 5. World Bank, Environmental Division, Africa Region, Washington, DC.

Faure, G., P. Dugue, and V. Beauval. 2004. *Le conseil à l'exploitation familiale. Expériences en Afrique de l'Ouest et du centre.* GRET-CIRAD, Paris.

Felix, A. 2006. "Eléments pour une refonte des politiques agricoles en Afrique subsaharienne." *Afrique contemporaine* 217: 159–72.

Griffon, M. 2006. *Nourrir la planète: pour une révolution doublement verte.* Odile Jacob, Paris.

Jouve, P. 2006. "Transition agraire: la croissance démographique, une opportunité ou une contrainte?" *Afrique contemporaine* 217 (1): 43–54.

Jouve, P., and M.,R. Mercoiret. 1987. "La recherche-développement: une démarche pour mettre les recherches sur les systèmes de production au service du développement rural." *Les Cahiers de Recherche-Développement* 16: 8–15.

Kleene, P., B. Sanogo, and G. Vierstra. 1989. "A partir de Fonsébougou. Présentation, objectifs et méthode du 'volet Fonsébougou' (1977–1987)." IER, Bamako.

Lavigne-Delville, P., N. E. Sellamna and M. Mathieu. 2000. *Les enquêtes participatives en débat. Ambitions, pratiques et enjeux.* Collection Economie et Développement. Karthala, Paris.

Lefort, J. 1988. "Innovation, technique et expérimentation en milieu paysan." *Les Cahiers de la Recherche Développement* 17: 1–10.

Legile, A. 1998. "Histoire de la gestion en France." Groupe de travail "outils et méthodes de gestion." Inter-Réseaux, Paris.

Lélé, U. 1989. "Population Pressure, the Environment and Agricultural Intensification: Variations on the Boserup Hypothesis." World Bank, Washington, DC.

Magrin, G. 2001. *Le sud du Tchad en mutation: des champs de coton aux sirènes de l'or noir.* CIRAD, SEPIA, Paris.

Mendras, H., and M. Forse. 1983. *Le changement social.* Armand Colin, Paris.

Mercoiret, M. R. 1994. *L'appui aux producteurs ruraux.* Karthala-Ministère de la Coopération, Paris.

———. 2006. "Les organisations paysannes et les politiques agricoles." *Afrique contemporaine* 217: 135–57.

Norman, D., and M. Collinson. 1985. "Farming Systems Research in Theory and Practice" in *Agricultural Systems Research for Developing Countries.* Richmond, Australia, May 12–15, 1985. *ACIAR* 11: 16–30.

Olivier de Sardan, J. P. 1993. "Une anthropologie de l'innovation est-elle possible?" Séminaire d'Economie Rurale "Innovation et societies." September 13–16, Montpellier.

———. 1998. *Anthropologie et développement. Essai en socio-anthropologie du changement social.* APAD-Karthala, Paris.

Opio-Odongo, J. 2000. "Roles and Challenges of Agricultural Extension in Africa." In *Innovative Extension Education in Africa,* ed. Steven A. Breth. Sasakawa Africa Association, Mexico City.

Pichot, J. P. 1996. "Diversité des systèmes de culture intertropicaux: un défi pour l'action." *Cah. Agric.* 5: 445–9.

———. 1998. Temps, espaces et acteurs : une approche globale de la durabilité des systèmes ruraux. *Oléagineux Corps gras Lipides* 5 (2): 104–5.

Pichot, J. P., M. Sedogo, and J-P. Deguine. 2006. "De nouveaux défis pour la recherche cotonnière dans un contexte difficile." In *Le coton, des futurs à construire. Cahiers de l'Agriculture* 15: 150–57.

Rivera, W. 2003. *Agricultural Extension, Rural Development and the Food Security Challenge.* Food and Agriculture Organization, Rome.

Schumpeter, J. 1935. *Théorie de révolution économique. Recherche sur le profit, l'intérêt et le cycle de la conjuncture.* Dalloz, Paris.

Triomphe, B. 1989. *Méthodes d'expérimentation agronomique en milieu paysan, approche bibliographique.* Mémoires et Travaux de 1'IRAT 19, IRAT, Montpellier.

Wampfler, B. 2006. "Innover et lever des tabous pour financer l'équipement agricole." *Grain de sel* 34 (35): 4041.

World Bank. 2006. *Enhancing Agricultural Innovation: How to Go beyond the Strengthening of Research Systems.* World Bank,Washington, DC.

Zoundi, J., K. Hussein, and L. Hitimana. 2006. "Libéralisation de la filière coton et innovation agricole en Afrique de l'Ouest." In *Le coton, des futurs à construire. Cahiers de l'Agriculture* 15: 17–22.

Chapter 8

Toward a Regional Food Market Priority

Anna Lipchitz, Claude Torre, and Philippe Chedanne

The food market is undergoing a profound change, marked by a sustained twofold increase: rising international prices and growing food demand in Africa. Examining the trends of the past 30 years is now necessary: low rates of protection, little public investment, targeting of exports (of niche or traditional commodities). Not only will the net import-export balance not drive growth, it could become negative.

To benefit from this new positive change, investment is required in the areas of intensification and infrastructure, standards-driven upgrades, vocational training, and the development of rural financing. It will be necessary, on the one hand, to focus on the growth of family farming operations, the parallel structuring of the agricultural sectors and producers' organizations, joint trade organizations, and commercial networks, and, on the other, to organize the market food sector on a regional market scale.

In Sub-Saharan Africa, the processes of agricultural transition appear to be blocked. Although the agricultural sector continues to employ the majority of the active population and 80 percent of the poor live in rural areas and are farmers, the share of agricultural gross domestic product remains minor, ranging from 33 percent for East Africa to 8 percent for Southern Africa. The share of African agriculture in world trade is negligible, but it continues to play a major role in feeding the African people. The first aim of this chapter, therefore, is to examine how Sub-Saharan Africa, which is in a weak position in the global agricultural markets, can become increasingly strong in the regional food markets.

New phenomena such as the increase in the prices of agricultural raw materials, the emergence of nontariff barriers, the signing of all-points bilateral agreements, and greater price volatility are dramatically changing perceptions of the role of agriculture on the markets. Meeting the challenge of African agriculture that is competitive on both the internal and the external markets makes it essential to analyze the manner in which the problems of marketing agricultural products arise. The second aim of this chapter is, therefore, to examine what role local marketing dynamics and agricultural market policies have in the context of restoring the full importance of the agricultural sector in economic growth and in the reduction of poverty and inequities.

Despite the Adoption of Liberally Inspired Models, the Share of African Agricultural Trade Remains Small

Scant Protection of Agriculture in Sub-Saharan Africa

Because the agricultural sector is viewed as secondary in most African policies, little thought seems to be given to the idea of providing direct support to protect and strengthen national and regional agricultural markets, in addition to which is the fear of losing the customs duties needed to finance budgets. For nearly 20 years, agricultural structural adjustment plans have thus engendered African national policies characterized by low customs tariffs; minimal direct intervention; the sudden and uncontrolled privatization of production, processing, and marketing enterprises; and the disappearance of stabilization funds and marketing boards.

The decline in customs tariffs among African countries and the adoption of common external tariffs (CETs) at levels well below the maximum rates authorized by the Agreement on Agriculture of the World Trade Organization (WTO) have both contributed to the low level of agricultural protection.

The tariff concessions required vis-à-vis Africa within the framework of WTO negotiations remain largely nonbinding. The first obligations in the multilateral context did not emerge until the Uruguay Round, which ended in 1994 with the Marrakech Agreement. Special and differential treatment exempts the least-advanced countries from reducing internal protections and supports. Twenty-one Sub-Saharan African countries benefit from this exemption. The other developing countries have limited, staggered reductions. At the same time, many countries are initiating bilateral liberalization processes, requiring them to lower their tariff barriers.

Sub-Saharan Africa's Minority Status on the Global Markets

In 2006, Sub-Saharan Africa as a whole remained a net exporter of agrifood products (figure 8.1).[1] Although its participation in the world trade system has increased in volume terms (exports have grown from less than $600 million in 1990 to $14 billion, and imports from $117 million to nearly $10 billion), its—still small—share of the global market has decreased. It accounted for less

Figure 8.1 **Trend of the Agrifood Trade: Sub-Saharan Africa and the Rest of the World**

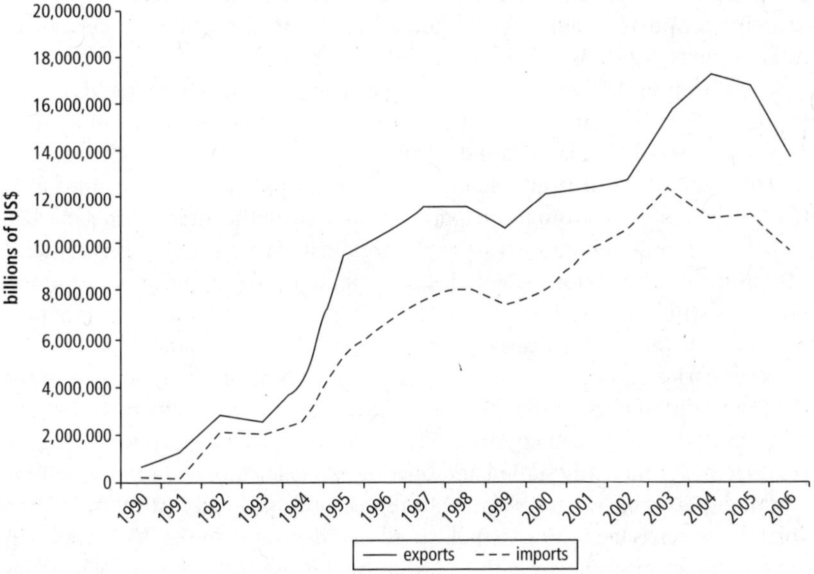

Source: Authors based on available data.

than 2 percent of total world exports in 2006 and only 1.4 percent of total imports. The share of family farming operations in total exports compared with agribusinesses is unknown and remains a matter of conjecture. Some products, such as pineapples, bananas, sugar, and flowers, are dominated by businesses with the capacity to invest, while others, such as coffee, cocoa, and cotton, are produced primarily on family farms. Moreover, imports also include food aid, because some African populations are subject to periods when grains are in short supply.

Agrifood exports, which account for 16 percent of Sub-Saharan Africa's total exports of goods, consist essentially of export crops, primarily coffee, tea, and cocoa (30 percent of agrifood exports), followed by fruits and vegetables (16 percent) and fish (14 percent). In 2006, nearly half of all exports went to the European Union, Sub-Saharan Africa's most important trading partner (especially the Netherlands, Great Britain, and France). Nevertheless, the situation varies widely from one country to another: agricultural exports can represent more than 80 percent of the exports of certain countries, such as São Tomé and Principe, Ethiopia, Burkina Faso, Malawi, the Gambia, and Benin.

Sub-Saharan Africa's agricultural imports are more diversified, including, in decreasing order, grains (24 percent of agrifood imports), oils and fats

(16 percent), and fish (10 percent). They represent 9 percent of the region's total imports of goods. For Eritrea, Niger, the Gambia, and São Tomé and Principe, imports of agricultural goods represent more than 30 percent of total imports of goods.

The European Union (primarily France, Spain, and the Netherlands) is the source of less than 30 percent of Sub-Saharan Africa's imports, followed by Argentina, Brazil, Thailand, and the United States.

The application of insufficiently protectionist policies has thus prevented the African countries from participating more fully in the international market.

Similarly, Sub-Saharan Africa's role in international negotiations remains insignificant. The region's countries participate in the adoption of common positions within the African group, consisting of 41 African countries that are members of the WTO (nearly a third of the WTO membership). The Sub-Saharan Africa countries are also part of the negotiating group known as the G-90, a coalition of African countries, African-Caribbean-Pacific (ACP) countries, and least-advanced countries. The African countries lack the means to put together sufficiently staffed and operational delegations to monitor all the working groups or to have permanent representatives in all institutions. Their voice, however, is beginning to make itself heard, such as in the WTO concerning cotton, for example, or with regard to the European Union in the negotiation of economic partnership agreements (EPAs).

Obstructed Intraregional Trade

Only 22 percent of the agricultural exports[2] of the Sub-Saharan Africa countries are bound for other countries in the region.[3] The available data indicate, however, that the regional market is the primary market for certain countries: for Gabon, Rwanda, Namibia, Swaziland, and Niger, regional exports account for more than 90 percent of total agricultural exports (figure 8.2).

Despite the adoption CETs, regional trade is still hobbled by numerous handicaps. As the implementation of the CET of the West African Economic and Monetary Union (WAEMU) attests, tariff dismantling has not been synonymous with trade growth in the African countries (Faivre Dupaigre 2007). Each country is indeed a unique case; the growth of its trade is explained more by its relations with its immediate neighbors and how it dealt with implementation of the CET. For many products (other than livestock and maize), the introduction of the CET seems to have had no effect in the WAEMU area. Indeed, multiple market failures prevent the advantages associated with the lower prices of imported regional products from being passed on to end users (industrial or consumer). The concentration of enterprises in a sector encourages profit motives more than diversification.

At the same time, government failures (racketeering systems, uncontrolled taxation, nonobservance of customs regulations), the weak business

Figure 8.2 Importance of Intraregional Agricultural Exports

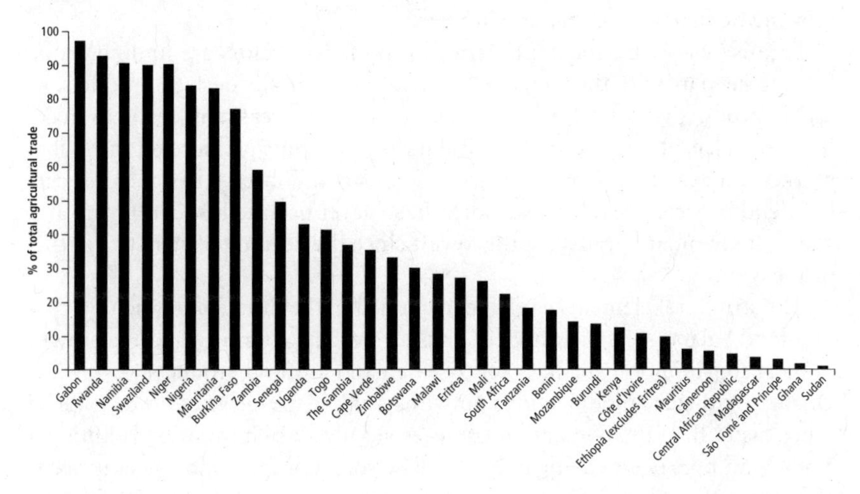

Source: COMTRADE.

environment (flouting of contractual rules, collusion between importers) and the myriad additional costs related to the condition of the infrastructure all act as constraints on investment and trade.

New Data Shake Up Conventional Thinking about Agriculture

Volatile Agricultural Commodities Prices

The economic development of emerging countries, changes in consumption patterns, and world population growth (1.2 percent a year) are increasing demand for agricultural products. The rise in meat consumption is also stimulating demand for grains and protein-oil crops to feed animals.[4]

In addition, the structural increase in the price of oil since 2004 and consideration of the negative externalities of traditional fuels (local pollution and global warming) are reviving interest in the agrifuel sectors. These sectors are steering the agricultural sector not only toward the production of foodstuffs but also toward the production of energy. Other nonfood markets are also influencing agricultural demand: generation of electricity or of nonmarket heating using wood, pure vegetable oil, methanization, or the thermal reclamation of agricultural wastes. Green chemistry is also becoming a nonfood market for agricultural products (production of bioplastics using starches, use of vegetable oils for motorization, solvents, and paints). Another factor is the growing demand for timber and lumber.

Because yields are stagnant, Africa's productive sectors are finding it difficult to keep up with the increased demand for naturally derived products. In highly productive countries and zones, gains are increasingly costly to make in technical, economic, and ecological terms. This puts pressure on the global markets and causes a structural decrease in stocks, observed as an accelerating trend in recent years; wheat stocks have never been so low in 30 years and stores of secondary grains are the lowest since the keeping of statistics in this field began.

The structural elements could be magnified in the future and weigh on current food balances. Higher incomes in the emerging countries and the growth of meat consumption are likely to stoke demand for food products worldwide. Qualitative dietary modifications notwithstanding, the curve of needs will rise much faster than the population curve, especially since most of the additional 3 billion human beings living in 2050 will be young people, whose energy needs are greater than those of older people. For Sub-Saharan Africa the problem is particularly troubling because of its burgeoning population and strong urbanization. Food aid and the financial resources generated by exports will not suffice, and the continent will have to boost its agricultural production more than fivefold by 2050 (Griffon 2006a).

Because of the rise in agricultural prices, and without any change in their competitiveness or population trends, developing countries were apparently faced with a 9 percent increase in their total spending on food imports in 2007 (10 percent for the least advanced countries). The basket of food imports of the least advanced countries in 2007 was reportedly 90 percent more expensive than the same basket in 2000. Other studies show that the caloric intake of poor populations falls approximately 0.5 percent whenever average prices increase by 1 percent. For Sub-Saharan Africa higher prices could have highly dissimilar short- and long-term effects, given the macroeconomic and microeconomic characteristics of each country in the region. At the macroeconomic level, not all Sub-Saharan African countries enjoy the same degree of food self-sufficiency and the same exchange rate for the U.S. dollar. In microeconomic terms, the short-term impact on agricultural households will depend on their "net buyer-net seller" position on the food markets. In the African countries, the international prices of foods are in fact "directors" of the domestic markets (that is, price-taker markets).

In addition, the prices of African agricultural exports such as cocoa, coffee, tea, and cotton have increased far more slowly in recent years than the price of grains. One of the most striking cases is cocoa, the price of which has stagnated since 2001, despite strong international demand. Moreover, in Côte d'Ivoire, the world's leading producer of cocoa, producer prices have been affected by the negative repercussions of mafia-style liberalization and the establishment of a jumble of opaquely managed structures.

All-Points Bilateral Negotiations

Although multilateral negotiations are at a standstill (owing essentially to disagreements in the agricultural sphere), bilateral agreements are proliferating and trade preferences—which govern trade relations between developing and industrial countries—are being overhauled. Trade preferences, however, must comply with the WTO legislation.

The most-favored-nation principle (which appears in the first article of the General Agreement on Tariffs and Trade) prohibits the granting of trade preferences. The "empowerment clause," which is the legal basis of the generalized system of preferences, makes it possible to circumvent this constraint by providing for more favorable treatment for developing countries. However, it prohibits any discrimination among developing countries not based on objective criteria. Thus, the preferential arrangements granted by the European Union to the ACP countries no longer qualify for the provisions of the empowerment clause, in contrast to the category of least developed countries, for example, which is defined by objective criteria. For this reason, the Lomé Convention arrangement, which currently governs trade relations between the European Union and the least developed countries, has benefited since its inception from exemptions granted by consensus of the members of the WTO (the most recent was granted with difficulty at the Doha conference and covered the period from March 1, 2000, to December 31, 2007).

Because this exempt arrangement may not be renewed, the European Union and the ACP countries have adopted the objective of moving from a system of preferences (nonreciprocal by nature) to a network of free trade zones (economic partnership agreements) encompassing the European Union and six regional groups. The negotiations, which were not concluded by the established time limit (the end of 2007), are continuing.

The consequences of the reciprocal tariff reductions associated with EPAs call for some caution. More specifically, a study (Fontagné, Laborde, and Mitaritonna 2007) shows the negative effects on the domestic meat, dairy, and wheat markets. Tax losses, without the implementation of an appropriate tax reform, could be prejudicial to public finances and the possibilities of financing and implementing meaningful agricultural policies. Will the continuation of negotiations lead to resolution? Provisions are definitely envisaged to limit the impact of unequal competition (sensitive products, accompanying policies), but there are other major challenges, such as the establishment of common external tariffs consistent with agricultural policies and the role assumed by new actors, such as Brazil, China, and India, relativizing the importance of the European Union as main trading partner.

Nevertheless, the proliferation of regional or bilateral agreements poses risks for African countries. Regardless of one's opinion of the WTO, it has the merit of imposing common rules and affording the opportunity to petition a dispute settlement body. These possibilities are largely circumvented within the framework of regional or bilateral agreements. In its report on trade and development released in September 2007, the UN Conference on Trade and Development shows that such agreements rarely yield long-term gains. By imposing restrictions on the ability of governments to promote long-term growth, these agreements can, in fact, thwart development objectives.

A Likely Increase in Price Volatility

Price volatility is linked to market mechanisms (more or less rapid and ample responses to price signals), unexpected shocks (climatic vagaries), and changes in the behavior of demand (short- and medium-term). Of particular note are same-year price fluctuations during a crop year, and multiyear changes. The effects of price volatility are of two kinds. At the macroeconomic level, countries are vulnerable to price volatility because it leads to variations in their export receipts, which can cause balance of payments crises. It also plays havoc with their tax revenues. At the microeconomic level, poor producers are the hardest hit, often selling off their crops on the markets at cheap prices. The decrease in their incomes considerably limits their future production capacities. The variability of their incomes limits their ability to invest and stifles innovation. It generates irreversible social effects. Finally, it can cause political instability and conflicts.

The phenomenon of price volatility affects developed and developing countries alike, but for the latter, whose markets are often more easily affected, the consequences can be more significant. Solutions to these risk factors have been sought in the past with the establishment of stabilization funds designed to smooth producer prices; others are being tested now (such as the "cotton smoothing fund" in Burkina Faso). The stabilization funds have been systematically dismantled over the course of the past decade, in parallel with deregulation and structural adjustment policies; this development has revived interest in seeking new tools to manage these price risks and their interrelationships (insurance, financial markets, national or international safety net).

Price volatility can increase owing to the very powerful repercussions of climatic events as well as to tariff concessions negotiated within the WTO or under bilateral agreements, with markets becoming less regulated and more open at the international level. This should lead to continuing examination of the advantages and disadvantages, on the one hand, of liberalizing agricultural markets and, on the other, of systems to minimize the fluctuation of agricultural prices. Because of the specificity of agricultural supply and food demand, does liberalization pose the risk of further instability and greater international disparities (Boussard, Gérard, and Piketty 2005)?

Nontariff Barriers, a Not Insignificant Constraint

Although the industrial countries' tariff barriers are tending to shrink and global wealth is increasing, compliance with sanitary and phytosanitary standards is considered the main obstacle to the participation of developing countries in international trade. The lack of harmonized international regulations can be a barrier to market entry (U.S., European, and Japanese markets for the most part, but emerging markets as well). At the WTO, the basic rules on the subject are contained in an agreement spelling out how governments can apply measures pertaining to food safety and health standards for plants and animals (sanitary and phytosanitary measures). These measures are designed to supply consumers with completely safe food products. At the same time, they aim to prevent strict health standards from serving as a pretext for the protection of national producers. These standards help reduce information asymmetries and transaction costs. They make it possible to incorporate an ethical and social dimension into products. They are also responsible for the creation of niches. In addition to public provisions,[5] there are private norms adopted by importers of agricultural products, large multinational enterprises, and major distributors. These norms—which concern not only the final quality of the product but also the production chain—create cost, infrastructure, and traceability problems for developing countries. Various studies have been conducted to assess the losses related to these norms and standards; the figures vary according to the sector, but the amount is considerable. The developing countries' access to a market can indeed be blocked and their market shares greatly reduced. However, the developing countries could use these norms as a catalyst to guide them in upgrading enterprises and promoting ranges of products.

A Priority, the Regional Food Market

Market Food Production as an Opportunity for African Agriculture

Heavy Demand, Together with Lucrative Prices. In Africa, population projections indicate a doubling of food demand between 2000 and 2015 (World Bank 2007). The urban population continues to grow and is expected to become the largest population group in the region by 2030. This trend in the urban-rural population ratio, with 80 percent of the rural population engaged in agriculture, explains why the market is now driven by heavy domestic demand. In West Africa, a network of cities is developing very rapidly within a band extending from the coastal countries to the Sahelian countries, putting pressure on the various markets (WALTPS 1998). This change will govern the trend of the market productivity of farmers, because production must be able to respond to demand growth of approximately 3 percent a year, representing an increase in marketable production of 5 percent a year (Cour 2004).

A second, new factor is the high level of "world" agricultural prices. This pressure on the markets is expected to persist in the short and medium terms, limiting for the African economies the supply of food products on the international markets. Ultimately, however, the situation could change with the arrival of second-generation biofuels, which are less competitive than food products.

The conditions for better producer incomes therefore seem to be met. This increased demand, together with sustained prices, is indeed a highly favorable factor for local production and greater returns on investments in the food sectors.

Supply That Has Kept Pace and Could Continue to Grow. The analysis of food production and consumption indexes as well as imports shows that the rural sector has largely succeeded in coping with a fast-growing population. Thus, between 1996 and 2003, four-fifths of the calories and protein consumed in Sub-Saharan Africa were supplied by national products (Bricas 2006). In West Africa, per capita food production has remained stable; in contrast, food imports grew from 10 percent to 16 percent of total imports between 1995 and 2004, partly owing to the rise in consumption, which climbed from 2,200 to 2,500 kilocalories per person per day. In addition, in that same region, the volume share of food production increased from 71 percent to 78 percent of total production between 1960 and 2005 because of the decreased interest in export crops (FAO and OECD 2007). Also in West Africa grain production grew threefold and the production of tubers grew fivefold between 1980 and 2006 (Blein, Faivre Dupaigre, and Soule 2008). The market value of African food production was $50 billion in 2003, compared with $16.6 billion for all agricultural exports (Diao, Dorosh, and Rahman 2007).

Regarding agricultural potential, major land resources are still available, especially in the wetlands (Blein, Faivre Dupaigre, and Soule 2008). The challenge will be to step up agricultural production in places where the potential for progress is enormous, even though extensive farming that destroys natural resources threatens to continue, owing to the pressure on land and migrations. Such intensification, carried out within a framework of sustainable ecosystem management and greater agricultural specialization, would likely be encouraged by the larger incomes earned by the richest categories of family farming operations.

Urban Demand, a Shaping Force. Although a large proportion of consumers are still rural, urban customers strongly influence the markets. The requirements of product quality and availability have created a new demand for services (processing, distribution, restaurant industry) linked to the urban lifestyle. Indeed, urban consumers are looking for products that are easy to cook, and they place a premium on identity (rural tradition, urban socialization) or economic (street consumption) references, with a particular attachment for products emphasizing typical features, naturalness, or innovation, which encourages the segmentation of food markets (Bricas 2006).

A More Accessible Technological Environment Made Possible by New Investors. In the past 10 years, China and, to a lesser extent, India have been increasingly involved in the agricultural equipment, agrifood, and vehicle markets in Africa. Manufactured equipment (motor pumps, mills, motorized cultivators, plows, seeders, motorbikes) is often quite sturdy, with available spare parts that can be repaired by small local shops. The price of this equipment is very low and substantially reduces the cost of the agricultural sector's access to upstream and downstream technologies (PASAL 2000).[6] The distribution of this equipment, through the granting of loans or subsidies, together with training, will be a highly effective tool for improving the productivity of the operations of all sectors.

A Window of Demographic Opportunity up to 2040. On the demographic front, reduction of the rate of dependence of the inactive[7] on the active population has already begun and will continue at a steady pace over the next 35 years. This suggests a demographic dividend made possible by the predominance of the employed population. If the latter work in growing economies, the prospective surpluses and savings could finance investments, particularly in rural areas. This opportunity is expected to start shrinking in 2040, owing to the growing number of dependent individuals over 60 years of age (Lipton 2006).

Market Food Production as a Regional Trade Priority

The growth rate of African food production (millet, sorghum, rice, roots and tubers, fruits, oilseeds, animal products) and the emergence of a sustained

BOX 8.1

Partial Conclusion

The most recent World Bank report (World Bank 2008) and the commitment of the African countries in July 2003 in Maputo to invest 10 percent of their budgets in the agricultural sector seem to indicate renewed interest in this sector after years of gradual abandonment. Today, a set of conditions favorable to the resumption of investment and improved productivity in the agricultural sector has emerged. However, 50 percent of the rural poor, while selling a large portion of their production, remain net purchasers of food products. It is therefore essential to ensure that their food security is guaranteed (Dorward, Kydd, and Poulton 2005; Lipton 2006).

Investing on a priority basis in the development of food markets can be a source of growth for both agricultural production and productivity, within a sustainable framework, while at the same time maintaining the food security of poor consumers.

demand with lucrative prices should encourage the effective implementation of policies designed to assist this movement with a view to promoting sustainable intensification, food security, and subregional trade.

Regional and International Markets. Without calling into question the relevance of the northern and far southern markets in African production, the subregional markets offer advantages that have not been sufficiently exploited. A simple, comparative exercise ignoring the interactions between the various crop systems makes possible the summary shown in table 8.1.[8]

Table 8.1 Summary Comparison of International and Regional Trade

	International export trade	Regional trade
Advantages and Opportunities	• Greater effectiveness linked to comparative advantages • Many East Asian countries currently developing based on this approach • Avoids the pitfalls of import substitution and promotes competition by eliminating rents • Facilitates qualitative upgrading of production techniques and provides a regulatory framework conducive to development • Access to foreign exchange	• Markets with unlimited potential, linked to the differential trend of urban-rural populations • Consumer arbitration facilitated in favor of regional products, furthering regional food sovereignty • Comparably developed regions with the same negotiating power • The value added created is captured by the region • Modest technological advances (no technical barriers to trade, sanitary, phytosanitary/threshold effect) • Does not require sizable exchange reserves • High energy prices = climatic consistency /change and a certain interest in the relocation of production activities
Disadvantages and Risks	• Growing regional and social inequities (Bienabe et al, 2006) • Poor domestic redistribution (Diao, Dorosh, and Rahman 2007) • Does not take account of threshold effects (institutions, level of development, capacities, regional economic assets, governance) • Requires substantial exchange reserves to import basic inputs • The rise in value of exported goods is not sufficient to jump-start growth (compared with rent-dependent countries) • Specialization can be impoverishing if it focuses only on primary products characterized by low demand elasticities relative to income • Markets dependent on world growth • Impact on climate change?	• Informal sector access difficult and process sometimes slow • Risk of creating ineffective regional rents through import substitution[a] • The prerequisites are good governance that eliminates corruption and the systematic protection of rents, transparency and account-ability, democracy, the promotion of an active civil society and policies on competition • Sustainability of production systems?

Source: Authors.

a. Economic history demonstrates, however, that many regions have developed under the umbrella of tariff protections and that good governance that promotes competition and reduces rents and that is free from political influence (transparency, accountability) is necessary to ensure the development of nascent industries.

If the largest possible number is aimed at, and if the local dynamics are given priority, the choice of market food production is obvious in the context of niche markets or traditional exports (Losch 2007). The relevance of this sector for economic growth and employment, compared with other types of agricultural production, is validated by the works of the International Food Policy Research Institute (Diao, Dorosh, and Rahman 2007) and the Latin American Center for Rural Development (Bienabe et al. 2006). This choice makes it possible to reach the mass of the rural poor and represents a model that considers development as occurring in different phases. Today, for Sub-Saharan Africa, a minimum of prior actions is necessary in the domestic agricultural markets before adopting a focus based more on exports or the manufacturing sector (Lipton 2006). This phase would correspond to the 1965–90 period in East and South Asia (Chang 2005; Ravallion 2008). Public intervention would serve to put in place the fundamentals and public goods (institutions, markets, infrastructures) before gradually withdrawing, leaving more opportunities for private investments (Dorward and Kydd 2005).

Market Food Production as an Instrument of Regional Integration If Trade Barriers Are Removed. Africa's regional food trade is the result of a long tradition of traders[9]—well known in West Africa (ENDA Diapol and CRDI, 2007), most often informal, organized into networks and solidarity chains—that enhance the complementarities between savanna (onion and tomato, livestock, grains) and forest (tubers, red oil, kola) agroecological systems, as well as between coastal and interior regions. The existing dynamics are very strong (FAO and OECD 2007) and should be used as a point of application of supports for regional integration, within the framework of public policies on regional land use.

Unfortunately, trade barriers are common in Africa (Alby-Flores et al. 2006) and represent market failures (lack of credit or insurance, information asymmetries, distortion of competition, immobility of production factors), government failures (abusive taxation, corruption, red tape), and trade constraints (deficient infrastructures, trade disparities, rules and regulations). These distortions explain the ineffectiveness of outward-looking policies that ignore supply capacities and the negative externalities affecting marketing circuits (hence the importance of marketing improvements by producer organizations, as indicated in box 8.2).

In these circumstances, African leaders should question the relevance of the choice of open regional spaces (Balié and Fouilleux 2005), especially when such opening is accentuated by monetary policies that can be highly unfavorable.[10] Remedying these weaknesses is therefore a priority for the development of these regions with a view to benefiting from economies of scale, the productivity gains resulting from the dissemination of technologies, and the investments

BOX 8.2

Producer Organizations and Local Marketing

In West Africa, countless producers try to sell their products themselves and have to contend with former field intermediaries or operate in a market dominated by merchants—circumstances that are highly disadvantageous for them.

It should be noted that producer organizations (POs) carry out and combine many activities to help producers overcome market access difficulties and enhance the value of their agricultural products. Numerous examples of studies within the framework of the Inter-Networks Working Group on PO initiatives in this area clearly show that POs provide services that enable their members—and, more broadly, producers in general—to access markets in more favorable conditions by adjusting balances of power, playing an intermediary role, facilitating negotiations, minimizing production risks, improving the quality and quantity of products, and lowering transaction costs.

The strategies are varied. Some are focused directly on improving production (quantity, quality) and capturing greater value added by "upgrading" the sector (vertical integration, transformation, facilitating direct sales to consumers). Others aim more at better control of market mechanisms: greater knowledge of prices and volumes offered, better control of market prices by matching supply and demand, creation of market spaces, improvement of the functioning of markets with regulations and control systems for greater transactional transparency, better management of supply and more comprehensive knowledge of demand.

So, should POs do everything? Can they do everything? Do they have the financial and human resources? Are they necessarily effective in doing everything possible to benefit member producers (processing, transport, marketing, research and experimentation, credit services)? What would be fair and effective in dividing responsibilities and activities between producers and POs, and between POs and other actors in the sector? If there is no unequivocal answer to these questions, which seems certain, the fact remains that POs are essential to facilitate the delivery of agricultural products to local markets and to provide missing services. By strengthening and regrouping themselves, to what extent will they succeed in winning over the national and regional markets?

Source: Anne Lothoré (Rural Development Inter-Networks), in Lothoré and Delmas (2009).

made possible by regional integration, all within the framework of community trade preferences.

Development that Permits the Sharing of Upstream Growth. Investment in favor of the agricultural food sector can be viewed as a pro-poor policy that makes the choice of targeting family farms and facilitates the promotion of employment,[11] the acquisition of assets (land, technology, equipment), and the expansion of food production (Lipton 2006). The expansion of production should be accompanied by increased factor productivity to reduce food prices and thus contribute to food security and the improvement of incomes (Dorward, Kydd, and Poulton 2005; Lipton 2006).[12] In this scenario, urbanization and the development of nonagricultural activities are increasing and serve as a stepping stone for future growth (Cour 2004; World Bank 2008).

A certain number of conditions are necessary, particularly in the equitable distribution of land (chapter 6 and Lipton 2006), innovation (chapter 7), financing (chapter 9), and incentive-creating macroeconomic policies (protective tariff policy or favorable monetary policy). Next, the evolutionary and progressive conduct of a policy of openness should make it possible to avoid rent creation by promoting increased productivity and access to technologies. Policies to reduce price volatility and therefore to improve food security will also be essential to boost production.

Moreover, because family farms are also the primary users of natural resources, they should be better compensated for their environmental services, to ensure the sustainability of those resources in a cooperative context.

Development that Permits the Sharing of Downstream Growth by Promoting Job Creation and Gender Issues. Although the supermarket revolution has barely begun in Africa, the strong growth of urban food demand has affected production owing to the dynamism of many different actors, most of whom are women in the informal sector engaged in the collection, storage, and processing of food products (Broutin and Bricas 2006). Although relying on traditional know-how, the artisanal sector has found innovative ways to gear its output to urban demand. The innovative capacities of these women operating downstream of the sectors make it possible to adapt a diversified demand for typical products to the requirements of standardization and presentation imposed by urban life. Technical innovations, mechanization and new processes have eliminated bottlenecks and enabled the sector to develop, as demonstrated by the growing urban consumption of attiéké, farinha/gari (cassava meal), cossettes d'igname (yam slices), and fonio and mango soups.

These micro and small informal agrifood enterprises represent the main sector of diversification of rural activities and account for a significant portion of gross national product (from 10 to 30 percent, depending on the country) and jobs. But very significant progress is still needed in the agrifood and trade sector, in terms of productivity, competition, infrastructure, and quality control and enhancement.

Generally speaking, intervention in the areas of infrastructure, equipment aid (transport, processing, storage) and credit availability, by helping to increase competition, should make it possible to reduce the subsectors' downstream costs and facilitate access to inexpensive food in urban areas, while at the same time maintaining remunerative prices for producers (PASAL 2000).

BOX 8.3

Partial Conclusion

Defined as an initial phase of development because it is focused on domestic rather than export markets, the development of food markets, with assistance, could become a powerful vector for improving the competitiveness of the various sectors and should make it possible to reduce poverty, while at the same time facilitating the emergence of institutions and actors that can intervene, in the medium term, in other geographies (export) or other sectors (manufacturing).

Requisite Conditions

Territorial and Sectoral Approaches Need to Be Made Consistent

Sectoral policies to promote domestic agricultural markets must be effectively formulated. The division between national and regional oversight authorities has not yet been clearly defined and often needs to be established, with efficient procedures and significant resources. Although the trade policies seem to be regional in nature, they still need to be actually applied. Most of the related national public policies (infrastructure, financing, and so forth) and the mechanisms for limiting price volatility (disaster funds, intervention funds, buffer stocks, microinsurance, early warning system, market information system) are not yet in place. The organization of markets (information, infrastructure, coordination, integration of sectors, storage) will be determining in this regard (Ministère des Affaires Etrangères–DGCID, GRET 2005).

No policy can be formulated or implemented without comanaging professional stakeholders, concerned about adjusting the balance of power among economic operators. In many cases, this cooperation between the profession and the public authorities is yet to be established. Institutional support and capacity building are essential to such an undertaking.

The sectoral approach makes it possible to identify limiting factors and, consequently, priority intervention points along the value chain from producer to consumer. Efficient tools (Dabat et al. 2008) are available to professionals and to the public authorities for performing diagnostic analyses and for better matching supply with demand. The inclusion of an uncomplicated monitoring and assessment system also makes available a decision-making tool, the outputs of which, clearly understandable, allow decision makers to strike a balance between more social priorities (employment), economic priorities (foreign exchange), and environmental priorities (impact on water or carbon emissions) (Kuper 2007). This tool supplements the macroeconomic modeling instruments.

At the sectoral level, the issue of harvesting is often considered the main limiting factor (situational rents) but also the one where the room for maneuver is greatest because credit favors competition between harvesters and "stationary merchants" (PASAL 2000). Interventions (research and development, followed by subsidies or loans) in the area of mechanized processing make it possible to remove bottlenecks (rice or fonio shellers and oil presses, for example) and to better distribute value added among the various actors (Broutin and Bricas 2006). The development of quality criteria (sanitary and phytosanitary measures, geographical indications, standards) make it possible to segment the market and better preserve value added in the production or processing stages. The reduction of postharvest losses also increases value added.

Strengthening the activities and the organization of the professional actors in a sector should enhance the competitiveness of products on the market and lead to greater consistency between upstream and downstream operations, more transparency, and a fairer distribution of value added. To that end, contractualization, interprofessional cooperation (box 8.4), professional structuring, and dialogue with the public authorities must be implemented without ideology; marketing can be handled equally well by competent professionals or organized farmers.

The organization of small producers and merchants into associations and cooperatives, as well as their personal training, should reduce transaction and marketing costs. The construction of this institutional infrastructure is essential in local approaches (Boselie and Van der Kop 2004); entry through local markets (identification of active players and products, sectoral diagnostics) makes it possible to combine a sectoral approach with the territorial approach.

Marketing assistance should focus on the monitoring of existing dynamics without attempting to create *ex nihilo* projects, which are costly and impermanent. This "market" approach allows for the development of a certain number of tools around market information, market price lists, storage resources, and warranted credit mechanisms, for example. Intelligent cooperation with local public authorities will permit the finely tuned management of territorial development. The development of markets for goods and services in rural-urban relationships will lead to the creation of numerous nonagricultural activities (Cour 2004).

Priority should be given to relying on local institutions, particularly microfinance institutions, which—because of their territorial organization, their cultural proximity to their "customers," and the skills they bring to bear—possess all the qualities of an effective "intermediary body." Support for the definition of financial tools geared to the varied needs of a target group of disparate actors will be essential. These institutions could also be used, by delegation of public authority, for the distribution of targeted subsidies. Such interventions will

BOX 8.4

Toward a Proliferation of Interprofessional Organizations?

Within the framework of transferring "responsibilities" from the public authorities to professionals, actors in the agricultural sectors have looked for new ways to manage relations among stakeholders and improve the performance of the sectors. Among the latter, interprofessional organizations are of particular interest. These organizations are composed of actors from at least two professional groups in a sector wishing to cooperate and jointly take action regarding a product. Organizations of this type have proliferated in Africa in recent years in both the export and the food sectors.

The first interprofessional organizations appeared in 1992 in Senegal with the creation of the National Interprofessional Groundnut Board (CNIA) and in Cameroon with the Interprofessional Coffee and Cocoa Board (CICC). Next were interprofessional organizations in the cotton sectors of several countries, established in the wake of the privatization of state ginning companies (Benin, Burkina Faso, Côte d'Ivoire, Senegal). Some countries (such as Burkina Faso and Senegal) have encouraged the creation of these organizations in the grain sectors. Many interprofessional organizations have also been established for different products such as dairy, bananas, tomatoes, poultry, and fisheries products.

Quite often, these groups take on multiple tasks, ranging from the promotion of local products to the establishment of trade agreements and the improvement of product quality, which neither the context nor their resources allow them to achieve.

The success of their activities will depend, among other things, on the level of organization and responsibility of their member organizations, their capacity to find sustainable and autonomous means of financing, and the degree of trust characterizing their relations with the state.

Source: Joël Teyssier (Rural Development Inter-Networks), in Grain de Sel (2008).

mobilize a network of training and advisory services—both local and regional, private and cooperative—to provide assistance.

National and Regional Agricultural Policies for the Penetration of Domestic Markets

Agricultural policies in Africa have often focused on exports in particular. However, new data on international agricultural trade reinforce the value of food products in relation to export crops. The long-term impact of the increase in prices will depend on the ability of agricultural producers to respond to price incentives. This presupposes the removal of several obstacles, specifically those that impede intraregional trade, such as the failure of a number of markets (credit, insurance, land) and the deficient economic capacities of the parties involved. If these obstacles are eliminated, the rise in prices will have an immediate effect owing to larger sales proceeds, which encourages the production of surpluses for the market and, therefore, the mobilization of production factors (land, labor, inputs, equipment).

To respond to the challenges outlined here, appropriate national and regional agricultural policies must be put in place. The regional dimension seems essential. In Africa, many regional entities have been formally established, but their proliferation and complex interrelationships make the actual implementation of regional trade difficult, and for certain regional organizations the problem is compounded by a limited budget.

Two policies are, nevertheless, worth mentioning in affirmation of the preponderant role of the agricultural sector in the countries' economies and development: the Agricultural Policy of the Union (APU), the first common agricultural policy approved by an African regional economic organization, established in December 2001 by the WAEMU; and the ECOWAS (Agricultural Policy of the Economic Community of West African States), which is the common agricultural policy of the Economic Community of West African States, made up of 15 West African countries, including the 8 WAEMU countries), adopted on January 19, 2005.

These agricultural policies must necessarily be the result of close cooperation among governments and the private and civil sectors (producers and consumers associations). Specifically, this requires strengthening the organization and expertise of the agricultural profession and of the rural actors.

Agricultural policies can meet the above-mentioned challenges only if they are sufficiently funded. At their meeting in Maputo in July 2003, the heads of state and government of the African Union therefore made specific commitments in favor of the agricultural sector, deciding "to allocate at least 10 percent of national investment budgets to the development of the agricultural sector in order to improve productivity and reduce food insecurity." This commitment was also reiterated in the draft common agricultural policy decision of ECOWAS. The share of public expenditure in this sphere in Sub-Saharan Africa is currently 4 percent. This decision seems to be having a greater effect than would a mere announcement—several countries have actually taken steps to invest in agriculture and are undertaking new commitments in the sector. Senegal, for example, has adopted a framework law for development of the agricultural-forestry-pasturage sector, as have Mali and Burkina Faso. Donors and lenders, which had themselves ignored the agricultural sector (the share of official development assistance allocated to the sector fell from 10 percent in 1990 to 4 percent in recent years), now view the agricultural sector as key to development.

These agricultural policies should address the problem of price volatility and, more generally, the issue of risk management. The combination of various tools could allay the destabilizing effect of sharp price fluctuations on the sectors, without, however, interfering with major market trends. A number of instruments should be developed and adapted to each risk segment. Again, the latter would be effective only if they are backed by strong farmer organizations and the firm commitment of the state and the international community.

Consistency of Public Policies

Agriculture will be able to fulfill its role completely as a contributor to growth and to the reduction of inequities only if the countries' agricultural and trade policies are consistent. This raises the question of the low level of external protection. New international trade data and these countries' population trends necessitate a reconsideration of such protection, which could be increased, at least up to the ceilings authorized by the WTO. In addition, there are numerous measures designed to maintain support for agriculture. Sensitive products can be identified and, as such, granted an exemption—if only temporarily—in the context of market liberalization. These products should reflect the productive incentives of the agricultural policy. There is no point in encouraging the production of grains, for example, if the applicable customs tariffs are excessively low. The identification of such products should also be based on the budgetary and tax contribution of each category of products. For the developing countries, other WTO measures illustrate how agricultural awareness can be taken into account (box 8.5).

It should be noted that these measures are largely underused. Policy makers and public policies do not take maximum advantage of this broad range of measures. Special products should also be clearly defined at the national level and then at the regional level. This requires arbitration, as yet nonexistent, to define common interests at both the national and the regional levels. This is another aspect of the debate concerning institutions and the role of public discourse. The special safeguard measure needs to be clearly defined. In addition to the possibility of developing countries activating it without conclusive evidence of the damage caused, it could also be initiated for price and volume factors more favorable to the developing countries.

Coordination of Sectoral Policies and International Cooperation

In conclusion, to restore the full importance of the Sub-Saharan African agricultural sector in economic growth and in the reduction of poverty and inequities, the industrial countries and the emerging countries must try to achieve consistency between sectoral policies and development policies. A first step has been taken in that direction: during the Sixth WTO Ministerial Conference in Hong Kong SAR, China, in December 2005, a decision was made to eliminate direct subsidies by 2013, as well as cotton subsidies by the end of 2006. However, the breakdown of negotiations at the WTO prevented elimination of the cotton subsidies. The European Union, with the reform of the common agricultural policy, is also committed to reducing its "distorting" agricultural support, by

BOX 8.5

Examples of Measures Designed to Protect Agriculture in the Developing Countries

Paragraph 13 on agriculture in the Doha Ministerial Declaration (November 2001) recognizes the development needs of the developing countries.[a] Among those measures, and within the context of market access, the package of July 2004 mentions provisions for special and differential treatment, related to Special Products and the Special Safeguard Mechanism, respectively, in paragraphs 41 and 42 of annex A (Framework for Establishing Modalities in Agriculture). The main objective of these provisions is to give developing countries greater flexibility in the area of market access.

Products are designated as special products based on criteria related to food security, livelihood security, and rural development needs. They will therefore receive particular attention in multilateral negotiations.

The current procedure of the special safeguard measure authorized by the WTO and officially available to all WTO members allows a country to impose an additional duty on a product subject to tariffs in the event of a specific increase in the volume of imports or a drop in the price of imports. However, this procedure is rarely used by developing countries. Some conditions are imposed on countries and on products. In addition, the trading partner must be convinced of the damage caused the national industry by the imports (a phenomenon that is difficult to prove without national legislation). Moreover, the special safeguard clause for agriculture can be invoked only for products for which tariffs have been established and provided the government has reserved the right to do so in its schedule of commitments concerning agriculture (only three Sub-Saharan African countries have reserved this right: Botswana, Namibia, and Swaziland). It cannot be invoked for imports entering the country under contingent tariffs. As a result, many developing countries, including African countries, which have not adopted the system of imposing tariffs on products have argued for the establishment of a special safeguard clause for agriculture.

The Agreement on Subsidies authorizes subsidies granted by developing countries whose gross national product per capita is less than $1,000.

a. "We agree that special and differential treatment for developing countries shall be an integral part of all elements of the negotiations and shall be embodied in the schedules of concessions and commitments and as appropriate in the rules and disciplines to be negotiated, so as to be operationally effective and to enable developing countries to effectively take account of their development needs, including food security and rural development."

delinking production from its assistance. Similarly, it is imposing a zero customs duty and free quotas for leas developed countries and is reducing the duties charged other developing countries. In Hong Kong SAR, China, the developed countries and the emerging countries agreed to do the same. This expanded access to developed-country markets is clearly a welcome development. In May 2007, the European Commission, bogged down in the economic partnership agreement negotiations, even decided to open up its market to all the ACP countries (except with respect to rice and sugar, which benefit from transition periods).

Several areas still require enhanced coordination at the international level:

- *Risk management.* A safety net, funded by the state, donors, and lenders, would make it possible to deal with catastrophic risks, the probability of occurrence of which is low.
- *Nontariff barriers.* The only rules and standards essential to preserve in the multilateral system are those that are necessary to protect public health, not those with a protectionist aim.
- *Trade assistance.* improvement of the countries' agricultural competitiveness will require not only the building of productive capacities but also the strengthening of their domestic infrastructures. In this context, trade assistance could serve as a promising new development tool. Representing 28 percent of official development assistance in 2005, trade assistance is to be increased, if the policy statements are to be believed. It could be used to support agricultural policies, stimulate the agricultural sectors to make them more competitive, facilitate intraregional trade through the establishment of infrastructures, and facilitate upgrades of productive resources. Its effectiveness is hampered considerably by the fact that donors and lenders currently have no trade assistance strategies in place. Numerous studies, however, give evidence of the search for better coordination of trade assistance projects.
- *Differentiation.* Recognition of the need for differentiation within developing countries.

Under WTO law, although special and differential treatment is supposed to benefit all developing countries equally and the status of developing country is self-proclaimed, the least developed country category clearly receives more favorable treatment, as do other, more narrow categories, such as vulnerable small economies. However, the agricultural sectors of other developing countries are particularly vulnerable and are in need of additional attention. This is the case, for example, of net food-importing developing countries and low-income food-deficit countries (LIFDC), as defined by the Food and Agriculture Organization based on objective criteria. These countries could be granted special status within the WTO.

Conclusion

Producing to sell is essential for any economic activity. In the case of agriculture in Africa, the emphasis has been placed on the importance of satisfying the local, national, and regional markets by giving family farms their opportunity first, because they must make a successful transition by benefiting initially from the closest markets; the issue in this case is to manage the trade-offs between farmers and rural and urban consumers, for whom this expenditure item is crucial. This does not, moreover, rule out relying on the strengthening of family farms and on agribusinesses to export and earn additional income and foreign exchange.

Agricultural raw materials are of vital importance in the commercial activities and in the internal trade networks of Sub-Saharan Africa. Consequently, greater emphasis should also be placed on the importance of the local markets, where women are key players; on the role of producer organizations in marketing and in the supplying of inputs; on new initiatives such as fair trade; on the influence of immigrant populations (Lebanese, Chinese, Indian); on the capacity of the powers that be to "spoil" the economic circuits and turn hidden rents to their advantage; and on the significance of storage, processing, and transport problems. In this part of the book, priority is given to an approach based on national and regional public policies and to promoting the search for consistency between agricultural, trade, fiscal, tax and monetary policies. Achieving consistency at the international level as well between official development assistance and agricultural, trade, and monetary policies is the next challenge.

Notes

1. Unless otherwise indicated, agrifood products in this text (whether processed or unprocessed) are understood according to the "MTN" classification: fisheries and fishery products, fruits, coffee, tea, cocoa, sugar, spices, grains, animals and animal products, flowers, plants, and beverages.
2. In this paragraph, fishery products are not included in the analysis.
3. By comparison, intra-European exports represent nearly 80 percent of total agricultural exports.
4. For example, 1 calorie of poultry meat requires the use of 3 calories of grain, and 1 calorie of beef, 7 calories of grain in nonpasture systems.
5. S standards, Regulations 852/2004 and 853/2004 comprising the health package, and 183/2005 for animal feed destined for the European Union.
6. Products subsidized in China by undervaluation of the yuan.
7. Population under 14 years of age and over 60.
8. For example, the fact that food production has increased substantially because of cotton production.
9. Such as Alahzaï, Dioula, and Peul merchants.
10. In macroeconomic terms, an overvalued currency is equivalent to low tariff protection and an export tax.
11. By promoting the capital-labor ratio, which is characteristic of small producers.
12. Poor producers, by producing more and buying locally, help to limit increases in food prices.

Bibliography

AFD, BMZ, DFIF, GTZ, KFW, and World Bank. 2006. *La croissance pro-pauvre dans les années 90: quels enseignements tirer de l'expérience de 14 pays?* "Operationalizing Pro-Poor Growth" research program.

Alby-Flores, V., B. Faivre Dupaigre, A. Vour'ch, and B. Yerima. 2006. *Accords de partenariat économique et dynamique des flux commerciaux.* Document de travail. AFD, Paris.

Azoulay, G. 2006. "Pour une sécurité alimentaire durable des pays les plus pauvres, quelques enjeux." In *Le monde peut-il nourrir tout le monde? Sécuriser l'alimentation de la planète.* QUAE and IRD, Paris.

Balié, J., and E. Fouilleux. 2005. "Une approche comparée des enjeux et processus de régionalisation des politiques agricoles en Europe et en Afrique." Colloque ASFP.

Berthelier, P., and A. Lipchitz. 2005. "Quel rôle joue l'agriculture dans la croissance et le développement?" *Revue Tiers Monde* 46 (183): 603–24.

Berthelier, P., A. Lipchitz, and N. Oulmane. 2003. "Quelles solutions pour dynamiser l'agriculture africaine?" *Diagnostics prévisions et analyses économiques* 26 (January) (http://www.minefi.gouv.fr/fonds_documentaire/Prevision/dpae/pdf/2004-002-25.pdf).

Bienabe, E., M. Fok, D. Sautier, and H. Vermeulen. 2006. *Case Studies of Agri-Processing and Contract Agriculture in Africa.* RIMISP Latin American Center for Rural Development (http://www.rimisp.org).

Blein, R., B. Faivre Dupaigre, and B.G. Soule. 2008. *Les potentialités agricoles de l'Afrique de l'Ouest (CEDEAO).* FARM.

Bonnet Grimoux, A., M. Buisson, H. Delorme, and A. Lipchitz. 2005. *Dynamics of International Agricultural Prices* (http://www.coordinationsud.org/article.php3?id_article=2116).

Boselie, D., and P. Van der Kop. 2004. *Institutional and Organisational Change in Agrifood Systems in Developing and Transitional Countries: Identifying Opportunities for Smallholders* (http://www.regoverningmarkets.org).

Boussard, J. M., F. Gérard, and M. G. Piketty. 2005. *Libéraliser l'agriculture mondiale? Théories, modèles et réalités.* CIRAD, QUAE, Paris.

Bricas, N. 2006. *Consommation urbaine et intégration régionale dans les pays ACP.* Colloque FARM.

Broutin, C., and N. Bricas. 2006. *Agroalimentaire et lutte contre la pauvreté en Afrique subsaharienne, le rôle des micro et petites entreprises.* Éditions du GRET, Paris.

Chang, H.-J. 2005. *Why Developing Countries Need Tariffs.* Oxfam-South Centre, Geneva.

Chedanne, P. 2005. "La crise des politiques agricoles en Afrique." *La lettre des économistes.* AFD, no. 10. Paris.

Cour, J.-M. 2004. "Peuplement, urbanisation et transformation de l'agriculture: un cadre d'analyse démo-économique et spatial." *Cahiers Agricultures* 13 (1): 158–65.

Dabat, M. H., P. Fabre, H. Hanak, and F. Lançon. 2008. *Manuel d'analyse de filières et logiciel.* CIRAD, Paris.

Diao, X., P. Dorosh, and S. M. Rahman. 2007. *Market Opportunities for African Agriculture.* IFPRI, Washington, DC.

Dorward, A. R., and J. G. Kydd. 2005. *Making Agricultural Market Systems Work for the Poor: Promoting Effective, Efficient and Accessible Coordination and Exchange.* Imperial College, London.

Dorward, A. R., J. G. Kydd and C. Poulton. 2005. *The Future of Small Farms: New Directions for Services, Institutions and Intermediation.* Conference on the Future of Small Farms. June. Wye, MD.

Dufumier, M. 2007. *Agricultures africaines et marché mondial.* Fondation Gabriel Péri.

ENDA Diapol and CRDI. 2007. *Les dynamiques transfrontalières en Afrique de l'Ouest.* Karthala, Paris.

Faivre Dupaigre, B. 2007. *APE et dynamique des flux régionaux: une application aux pays de la CEDEAO.* AFD, Paris.

FAO (Food and Agriculture Organization). 2007. *Perspectives de l'alimentation.* FAO, Rome.

FAO and OECD (Organisation for Economic Co-operation and Development), Coordination J. Bonnal. 2007. *Les ruralités en mouvements en Afrique de l'Ouest.* Conférence internationale sur la réforme agraire et le développement rural (CIRADR). (http://www.fao.org/docrep/fao/010/ah835f/ah835f.pdf).

Félix, A. 2006. "Éléments pour une refonte des politiques agricoles en Afrique subsaharienne." *Afrique contemporaine* 217.

Fontagné, L., D. Laborde, and C. Mitaritonna. 2007. *Étude de l'impact des APE et de l'intégration régionale sur les pays ACP.* CEPII. Presented at the CEPII conference, September 19, Paris.

Grain de Sel. 2008. "Les organisations inter professionnelles, des outils pour l'avenir des filières." Grain de Sel 44.

Griffon, M. 2006a. "Les agricultures du Sud et les responsabilités européennes." In *Le monde peut-il nourrir tout le monde? Sécuriser l'alimentation de la planète.* QUAE and IRD, Paris.

Griffon, M. 2006b. *Nourrir la planète.* Odile Jacob, Paris.

Groupe de pilotage GREMA. 2006. *La régulation des marchés internationaux de produits agricoles.* Pratiques de la régulation des marchés agricoles internationaux et outils de régulation et stratégies des acteurs (http://www.coordinationsud.org/article.php3 ?id_article=2449&var_recherche=gemdev).

Josserand, H. 2006. "Insécurité alimentaire dans le monde. Implications pour les pays européens." In *Le monde peut-il nourrir tout le monde? Sécuriser l'alimentation de la planète.* QUAE and IRD, Paris.

Kuper, A. 2007. *Élaboration des politiques pour les filières agroalimentaires, capitalisation des travaux de recherche et études, données, outils et méthodes.* Rapport principal CGDA, Morocco.

Le Coq, J.-F., and V. Ribier. 2007. "Renforcer les politiques publiques en Afrique de l'Ouest et du Centre: pourquoi et comment?" *Notes et etudes économiques* 28.

Lipchitz, A. 2007. "Les accords de partenariat économique: des accompagnements nécessaires." Document de travail 36, AFD, Paris.

Lipton, M. 2006. *Can Small Farms Survive, Prosper, or Be the Key Channel to Cut Mass Poverty?* Presentation to FAO Symposium on Agricultural Commercialisation and the Small Farmer. Rome, May 4–5, 2005. *E-Journal of Agricultural and Development Economics 3* (1): 58–85.

Losch, B. 2005. *Structural Implications of Liberalization on Agriculture and Rural Development*. Concept note, revised version.

———. 2007. *How Can Rural Producers Become More Competitive in the Face of Globalization and Supply Chain Integration?* Rural Structure Program. 2nd European Forum on Sustainable Rural Development. Berlin, June.

Lothoré, A., and P. Delmas, ed. 2009. "Accès au marché et commercialisation de produits agricoles: Valorisation d'initiatives de producteurs." Inter-réseaux Développement rural, AFD-CTA, Paris.

Mazoyer, M., and L. Roudard. 2006. *La fracture agricole et alimentaire mondiale.* Universalis, Paris.

Ministère des Affaires Etrangères–DGCID, GRET. 2005. *L'appui à la commercialisation en milieu rural,* l'actualité des services aux enterprises 9 BDS (Business Development Services). (http://www.gret.org/ressource/bds.asp).

PASAL, DYNAFIV. 2000. *Rapports d'activités du projet d'appui à la sécurité alimentaire.* Ministère de l'Agriculture, bureau de coordination des politiques agricoles. Guinée.

Ravallion, M. 2008. "Are There Lessons for Africa from China's Success Against Poverty?" OECD-DAC POVNET Workshop on China and Africa, Paris.

United Nations Conference on Trade and Development. 2007. *Rapport sur le commerce et le développement 2007.*

WALTPS, Club du Sahel. 1998. *Pour préparer l'avenir de l'Afrique de l'Ouest: une vision à l'horizon 2020.* Under the direction of Cour and Snrech. (http://www.oecd.org/dataoecd/50/16/38513077.pdf).

World Bank. 2007a. *L'agriculture au service du développement.* Rapport sur le développement dans le monde 2008, World Bank, Washington, DC.

———. 2007b. *World Development Report 2008: Agriculture for Development.* World Bank, Washington, DC.

Chapter 9

Financing Agricultural and Rural Transitions

François Doligez, Jean-Pierre Lemelle, Cécile Lapenu, and Betty Wampfler

At the outset, the financing of African agriculture was based on a public economy system that was broadly called into question in the course of structural adjustment, although the private sector did not prove itself capable of taking its place to the extent necessary and with sufficient diversity. To be sure, new categories of stakeholders—in particular in the microfinance sector—as well as diverse methodological innovations did emerge and make it possible to provide some responses, but these remain broadly insufficient. There is thus a need to explore new and supplementary paths, based on financial systems or on production, paths that steer clear of the utopian vision of a single solution that is applicable everywhere. Such innovative approaches must be part and parcel of the broader political and economic context governing agricultural transitions.

To accommodate the transformation of family agricultural systems, the matter of financing—its origin as well as the intermediation paths toward farmers and their organizations—is a crucial issue regarding the agricultural challenges analyzed in this study. Indeed, to improve the productivity of small farms and strengthen the rural economic fabric, better access to financial services—savings, credit, investment, or insurance—would appear to be a critical factor in the agricultural transition. While there is nothing new about this area of concern, the manner of approaching it has evolved considerably since the end of colonialism in Sub-Saharan Africa as in all countries of the South. This chapter addresses the historical background of agricultural finance in West Africa and raises questions about innovations that need to be introduced regarding agricultural financing as well as the terms for implementing such financing.

Historical Background of Agricultural Financing and Its Limits in West Africa

A Look Backward

From independence until the 1980s, the financing of agriculture was considered primarily to be the province of the state. The scant presence of private financial institutions and the absence of available resources justified the creation of development banks in most African countries. Their efforts were supplemented by market support and regulation mechanisms in the form of public support and price equalization organizations that guaranteed prices, as well as direct interventions through budget instruments, particularly in project form. The banks were called upon to finance all development programs and projects, in particular in the areas of infrastructure, industry, crafts, and agriculture, with a view to achieving "accelerated modernization." However, as in many countries of the South, the official agricultural financing systems of Sub-Saharan Africa—specialized banking institutions or public agencies—experienced severe crises in the 1980s as a consequence of management weaknesses. Triggered by the spike in interest rates on the international financial market and the growing insolvency of the indebted countries, the crisis was reflected either in the bankruptcy of those institutions or in their scaling back into quite specific niches (state corporations, import-export, export zones and crops, hydroagricultural improvements), leaving the majority of rural economic stakeholders, family farms in particular, outside any institutional financial circuit.

This was the case of the member countries of the West African Economic and Monetary Union (WAEMU), where four of the seven agricultural development banks or national agricultural credit funds were liquidated (Togo, Niger, Côte d'Ivoire, and Benin), one experienced severe financial difficulties (Senegal), and two of the more recent ones (Mali, Burkina Faso) retrenched to financing cotton sectors that were guaranteed by the marketing monopoly of the state corporation (Le Breton 1989). In the context of their structural adjustment programs, the African countries, like many other developing countries, were constrained in the 1980–2000 period to reduce their financial imbalances and therefore to disengage from credit systems that were often poorly managed and highly costly, owing to the poor repayment rates. The financial deregulation espoused by the international institutions was intended to reduce the role of the state, thereby eliminating the distortions associated with a managed economy, and was expected, thanks to rising interest rates, to make it possible to increase national savings and thereby revitalize investment without having recourse to external borrowing (World Bank 1989).

This deregulation of financial markets was not accompanied by improved coverage of demand. On the contrary, the majority of economic stakeholders remained cut off from access to credit. During this period, the supply of credit was assured through:

- local initiatives, *tontines* to private lenders, which address some needs for mutual assistance, solidarity, or emergencies. These are sometimes quite dynamic, as in the case of the itinerant bankers who, in southern Benin via Nigeria or Ghana, gather savings from small merchants on the overnight markets and transform them into credit (*Revue Tiers Monde* 1996). In general, however, they were not sufficient to meet needs owing to their limited resources and the cost of credit from them.
- the reform of public banks, allowing for the granting of lines of credit through agencies, or subsidies for the purchase of such things as fertilizers and equipment by small farmers.
- the emergence and development of an intermediary sector, between local initiatives and the banking system, which permitted both rural and urban access to financial services and was strengthened with support from public international cooperation groups such as nongovernmental organizations (NGOs) (Gentil and Fournier 1993). Inspired by the cooperatives existing in Europe since the end of the 19th century or the solidarity credit popularized by Grameen Bank of Bangladesh created in the 1970s (Yunus 1997), this new microfinance sector took off rapidly in the 1990s (Lelart 2005) and has gradually been further structured through special national provisions and legislation (Lhériau 2005). Against the backdrop of global economic crisis during the "lost decade" in Sub-Saharan Africa, there are only rare sectors that have experienced such growth (table 9.1).
- other modes of supplementary financing that have gradually been structured, namely:
  - financing by sector; commercial banks have been permitted to reach out to producer groups and get involved in supplying and marketing certain products (cotton in particular)
  - the financing of rural communities—through local development funds and projects and subsequently through decentralization arrangements. This instrument has grown in importance in the financing of infrastructure and in numerous other investments with indirect profitability that lending cannot cover.

Supply and Demand in 2006–07

Today, the economic stakeholders excluded from the banking sector, such as family farmers, craftsmen, or merchants, a sizable proportion of whom are often women, see their financing needs on the rise with the opening of markets and the increasing monetarization of their commercial activity.

Table 9.1 Trend of Savings and Loan Institutions in West Africa

	1993	1995	1997	1999	2001	2003
Number of institutions	107	174	189	272	303	—
Number of members (1,000s)	466	743	1,441	2,352	2,789	3,146

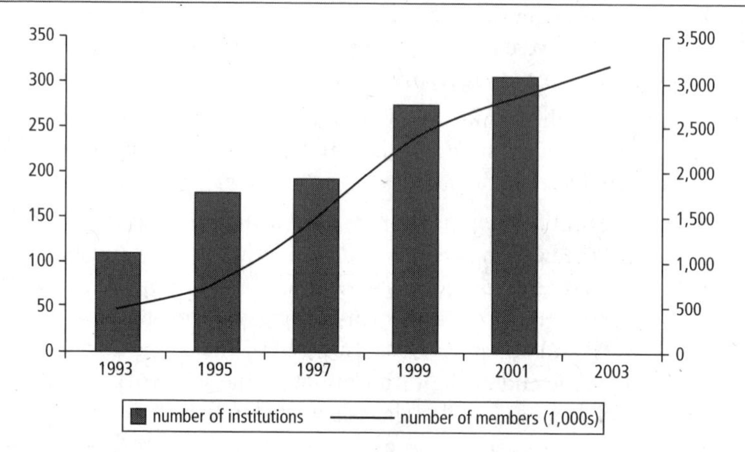

Source: PASMEC-BCEAO, 2000, 2002, and September 2003.
— : Not available.

Contrary to the myth of self-subsistence, small farmer economies have been integrated into the market for quite some time. Even the farmers in the most isolated zones need to purchase a portion of their consumer goods, and increasingly often part of their revenue-producing goods, to which end they sell agricultural products or their labor.

In response to the seasonality of incomes related to the agricultural cycle, but also in response to unpredictable and unforeseen developments that bear on the most vulnerable economies, access to credit has become a necessity for financing agricultural production, as well as for the survival of many rural families (Servet 2006). Finally, the capitalization of productive systems is often necessary in order to achieve productivity gains.

Despite the pronounced surge in new mechanisms in recent years, demand is far from being covered (Joseph 2000). In terms of access to financial services on the part of farmers and other rural economic stakeholders, microfinance falls short of meeting needs. In many countries, the "reach" of financial services remains quite limited, far from the performance level expected to ensue from financial deregulation (table 9.2).

Table 9.2 Reach of Selected African Financial Systems
Percent of adults

Category	Zambia	Botswana	Namibia	South Africa	WAEMU
Users of bank services	14.6	43.2	51.1	47.0	3.0
Financially included	7.8	5.8	2.0	8.0	15.0
Access to informal mechanisms	11.3	5.0	2.7	8.0	—
Excluded	66.3	46.0	45.2	37.0	—

Source: Various authors, *Techniques FINANCIERES et DEVELOPPEMENT* 84 (September 2006).
— : Not available.

This problem is accentuated in the countries where agriculture accounts for an important share of the economy and employment (Morvant-Roux and Servet 2007). Paradoxically, signs of saturation of the urban microcredit markets are appearing in various countries, including some in Africa (Azokli and Adjibi 2007).

Many already well-known factors make it difficult for family farms and rural economic units to access financial services, in particular:

- the territorial dispersion of borrowers, distance from places of residence, isolation of some regions, and low population densities in many rural situations, which increase financial services transaction costs
- levels of income poverty that are often high, and, in addition, the small size of the unit amounts to be managed for the various transactions (deposits and loans), which are often not profitable given the unit costs of the individual records that must be borne by the financial intermediaries
- the high risk levels and extremely diverse nature of risks (climatic, economic, social), which are often covariant within one and the same region
- the weakness of guarantees that can be readily realized from the legal standpoint (mortgages or pledges, for example), or from a social or economic standpoint (precarious living conditions or livelihoods)
- the weakness of human capital available locally (illiteracy, difficult living conditions, limited access to services), which increases the management constraints on institutions owing to the problems and costs of recruiting personnel locally or making employees available from other residential locations, particularly urban ones
- financial discipline, often referred to as the "culture of credit," where a loan is sometimes treated as if it were a grant—or an entitlement—owing to the relationships between the authorities and rural populations, because of numerous institutional antecedents in the area of credit, which still linger in various social or technical approaches, during electoral periods, or in clientelist-type policies

Faced with the stakes at play in agricultural and rural transitions, improving the financing of agriculture and rural economies remains a priority and a sine qua non precondition. It means that new orientations and innovations need to be proposed, which are based, on the one hand, on instruments that have gradually been structured in the recent period, and, on the other hand, on new approaches to be tried out.

How Can the Financing of Agriculture and the Rural Economy Be Improved?

Meeting the Needs of Farmers and Their Organizations

To achieve adaptation and transformation, family farmers have needs for technical and organizational innovations that cannot be covered solely by their capacity for self-financing. Their financing demands are sizable, diversified, and complex. A review of the needs of family farming was sketched out at a meeting in Dakar in 2002 (CIRAD-CERISE 2002). It may be revisited and supplemented by showing the following needs for the financing of family agriculture:

- short-term: financing of inputs at the start of the crop season (seeds, fertilizers, pesticides) and of supplementary labor; renting or share-cropping; livestock fattening, storage in order to take price shifts into account, processing of production to maximize value, diversification of income-generating activities
- medium- and long-term: equipment for intensification, marketing (transport), storage (buildings), perennial crops (investment, renewal, maintenance), rebuilding herds, the purchase of land
- family needs: personnel, equipment, housing, education
- savings to address needs in the various cycles (agricultural seasonality, investment, life cycle), but also as provision for the future or protection against unforeseen developments
- insurance services for the risks associated with family health and access to care, material needs, agricultural and livestock production, natural disasters, and climatic risks
- nonfinancial services: technical support and advice, management assistance, marketing support

With agricultural transitions toward more productive farming approaches and with the strengthening of the economic organizations that support them, new needs emerge. These pertain to:

- agricultural "entrepreneurs" who have significant needs for the financing of their activities and their investments. Access to financing—supplementing their own self-financing—is a decisive factor in determining the pace and quality of the development of their agricultural "business." They constitute a financial market segment that needs to be built up (see the comments on mesofinance below).

- agricultural professional organizations that need prefinancing of stocks of inputs, rolling capital for marketing activities, equipment needs, buildings, and so forth. They may also work to support the financing of their members.

Innovation in Products and Services

Innovations are now well identified and seem to be either already taken up or in the process of validation, through a number of significant experiments conducted by rural microfinance institutions.

- Leasing, inspired by hire-purchase operations, is an alternative to conventional medium-term financing for equipment, which makes it possible to eliminate the constraint of collateral. Leasing consists in separating the ownership of an asset from the usufruct thereof: the institution remains the legal owner of the equipment until the customer has completed repaying his hire-purchase agreement. Tried with some success within projects to promote work with draft animals, in particular in Madagascar, leasing has been adopted and streamlined by numerous microfinance institutions.
- "Storage credit" or agricultural warrantage is aimed at securing credit for farmers, based on contracts for the storage of their harvests. This enables producers to use this form of credit as security for a loan intended to cover the costs of marketing or processing, to develop income-generating activities between seasons, or wait for international prices to rebound. Various institutions, in Madagascar and Niger in particular, are developing "common village-level granaries" so that producers can obtain the best possible return on their production while waiting for the gap between crop seasons to send it to market.

Other innovations are being made and extended to other types of financial institutions seeking to expand their services in rural areas.

- Mutual surety associations and companies can support the "upscaling" of rural credit and compensate for the absence of collateral on the part of small entrepreneurs. Often established within a professional network, they enable financial institutions to grant much larger loans in order to boost the dynamism of some subsectors (agricultural harvesters in Guinea, craft workers in Burkina Faso).
- To reach scattered rural populations, so-called "branchless banking" or "mobile banking" systems are being tested for offering financial services

outside the conventional framework of bank branches. They often bring together a financial institution (a bank or microfinance institution), an operator of new technology (electronic payment terminal, Internet server, mobile telephony), and a retailer (merchant, NGO, or post office).

Finally, new innovations are currently the subject of experimentation, both in the insurance area and in mesofinance.

In conventional crop insurance, any damage to crops must be verified in the field before compensation is paid out. However, damage assessments are expensive, and precisely measuring the loss at each insured farm is even more costly. To reduce these costs, indexed policy systems that work differently are being tried out. A given meteorological measure (temperature, precipitation, wind, lack of rainfall) is used as a factor triggering payment of the indemnification. The insurance policy is replaced by a coupon that becomes payable whenever the meteorological event occurs. But this kind of indexed insurance has its limits (Autrepart 2007). It is therefore still a question of drawing lessons from both traditional agricultural insurance and indexed insurance with a view to protecting farmers more efficiently.

The modernization under way in the rural environment is fostering the development of investment, which is one of the vectors for improving productivity, product quality, the possible processing of products, and marketing, and this has given rise to a new approach: mesofinance. The term *mesofinance* is used to designate one segment of financial need included within a broad bracket from €2,000 to €100,000, which is covered neither by microfinance nor by the local banking market (box 9.1).

Although a number of promising innovations have been developed for the financing of agriculture, the generalized application of these techniques is still problem laden. It is likely that the communication and information-sharing between institutions was too limited for a long time, but these innovations are now becoming better known and examples of their application are disseminated ever more widely. A number of technical constraints also limit these innovations: the financial resources remain inappropriate, the regulatory frameworks are often too restrictive, and the skills of the local stakeholders are still limited. Sometimes, it is also observed that these innovations remain too costly for the beneficiaries. Finally, financial services to agriculture cannot be effective unless they are part and parcel of an active rural economy that is supported by functional services: the provision of inputs, marketing, and agricultural and rural advisory services for improving production and management techniques, information systems on markets. Thus, one of the success factors relates to the partnerships that can be forged between financial stakeholders and other service providers to support borrowers.

BOX 9.1

Outlook for Mesofinance

Today, there are multiple obstacles to the development of mesofinance. Among them, mention may be made of insufficient long resources, interest rates, and the absence of collateral on the part of the microfinance institutions as well as the poor understanding of this customer segment, the increase in transaction costs, the lack of collateral, and the lack of know-how on the part of the banks.

Accordingly, programs supporting the development of mesofinance are in the process of being developed, rural areas included. The programs tend to support at the same time the supply of financing (banks) and the demand (agricultural entrepreneurs). On the supply side, this means supporting the banks in "downscaling" and targeting a new customer group by providing technical assistance to overhaul methods as well as a partial portfolio guarantee to share some of the risks relating to this new loan portfolio segment. On the demand side, the programs endeavor to provide support of capacity building of the enterprises themselves (training in management and accounting, technical training, and so forth), but also of their representative bodies so that these can ultimately take over the function of providing nonfinancial services for their members.

Finally, the programs supporting mesofinance can support the structuring of mutual surety companies operated by the professional associations. This support, geared to raising the barrier associated with the absence of any real guarantee, essentially involves providing the start-up guarantee and support for setting up governance modes, which are often complex.

Source: Aude Penent (Micro-Finance Project Manager, AFD).

Promoting Multiple Approaches: Stakeholders-Institutions-Subsectors

An initial approach for rural and agriculture finance uses the financial sector as the point of departure and is articulated around the financial institutions to facilitate access to a wide range of services. Various discussions are structuring this approach to rural and agricultural finance by the financial sector, in particular focusing on the forms of organization, the scale and distance, and on links with urban finance. There is often a continuum between rural finance and urban finance, but a trend has been observed on the part of rural institutions to aspire to growing toward the urban environment, which is considered to be more profitable and capable of ensuring the viability of the organization. A price equalization system or solidarity system is therefore put in place to ensure the overall profitability of organizations through profitable urban activities. However, a number of questions arise, first and foremost regarding the drifting of microfinance institutions away from their mission of financing the rural sector. In addition to the opportunities for making loans in higher unit amounts, their branching out into urban areas is also aimed at lowering the cost of the resource by collecting savings, but this entails specific risks.

The objective of this financial approach is to build long-term capacities and to identify incentives that enable institutions to offer appropriate financial services to the rural and agricultural sector (box 9.2). This approach has the advantage of enabling a broad range of services to be organized, including medium-term loans, savings, insurance, and transfers. It also facilitates access to external resources and is less dependent on agricultural monospecialization and on the economic conditions within a single production subsector.

BOX 9.2

The Principal Organizational Forms of Agricultural and Rural Finance

Savings and loan cooperatives are the primary component of rural microfinance. They are managed by their members, with support from wage-earning personnel. Within African microfinance, the mutual networks make the strongest contribution to financing agriculture. The cooperative and mutual networks grouped together within the Center for Financial Innovation (*Centre d'innovation financière,* CIF), such as Kafo Jiginew in Mali or the Mutual Agricultural Savings and Loan Funds (*Caisse d'épargne et de crédit agricole mutuel,* CECAM) in Madagascar, have been able to offer adapted services for the financing of agriculture, including medium-term credit for equipping farms when, in particular, there are governance arrangements bringing agricultural producers together. In response to the difficulty of mobilizing savings, especially in rural areas, numerous adaptations of the basic principle of prior savings and cooperative models may now be observed.

The models using solidarity surety, of the Grameen Bank type, call upon the social cohesion of a group of five to ten persons to guarantee repayments. Financial transactions are carried out through credit officers. The strict model (small cohesive groups and wage-earning credit officers) is poorly adapted in rural areas, particularly in Africa, because of the low population density, high transaction costs for groups, covariant risks on activities that might create pressures within groups and make the principle of solidarity surety ineffective, and migrations out of rural areas that are incompatible with the mode of operation based on regular meetings. In rural areas, it is rather the participatory models—cooperatives, village-level associations, and the like, in which the members assume responsibility for a portion of the transactions in the place of the credit officers—that play a more important role.

The Self-Managed Village-Level Savings and Loan Funds (*Caisse villageoises d'épargne crédit autogérées,* CVECA), such as the village-level banks supported by NGOs, have been developed to make it possible to deliver profitable savings and loan services in rural

(continued)

BOX 9.2

(continued)

areas with low population densities, in particular to address the needs of the Sahelian zones. They are based on the possibilities for mobilizing local savings (household capacity, attractiveness, culture) supported by refinancing lines, and on the existence of community solidarity and self-management capacities (or wills), but they generally are not heavily involved in financing agriculture resource constraints, small units more sensitive to covariant risks).

The commercial banks are also getting more and more involved in rural zones by investing in local financial institutions, by establishing branches, or by proposing refinancing lines to microfinance institutions active in rural zones. These relationships are based on a long-term partnership and geographical proximity: it is often the individual country branches that refinance the microfinance institutions.

More recently, the development banks have regained currency, in particular in Latin America and Asia, where they underwent reform without disappearing with financial deregulation. In international guidance, one can observe a reorientation toward a "rehabilitation" of public agricultural banks, which can create, thanks to new forms of governance, public-private partnerships that respond to the needs for rural and agricultural finance. Nevertheless, it remains to be seen whether the new generation of solidarity banks or agricultural development programs will be able to draw the proper lessons from the earlier failures or bankruptcies of the agricultural banks in the 1970–80 period.

A second approach to rural finance is focused on the subsector, or "value chain." It uses the production subsector as the point of departure and structures the proposed financing all the way down the agriculture value chain (through, for example, suppliers of inputs, processors, intermediaries, or buyers). The financial services are most often combined with marketing activities and possibly with technical assistance.

The approach of financing through agricultural subsectors is aimed at reducing the risk of nonrepayment. This financing method has long been the main vector of financing for farms in various export subsectors (cotton, cocoa, coffee). Similarly, some production subsectors are sometimes associated with contractual agriculture organized by the distribution system (milk in Madagascar, large-scale distribution in the case of vegetables in Kenya, fair trade in various products) and allows for access to financing coupled with associated services (technical support, training, contractualization of market outlets).

Here, the comparative advantages of the subsectors—as compared to the financial institutions—make it possible to avoid the information constraints between stakeholders and facilitate the acceptance of nontraditional forms of guarantees, such as standing crops or stocks, thus resulting in better secured financial services. Their management costs are also reduced by "integrated" repayment mechanisms. Nevertheless, the financing is tied to a specific production subject to fluctuations in price and market outlets. The frequent monopoly situations may imbalance the sharing of costs and risks to the detriment of the farmers.

Sometimes, a more balanced approach involving combinations of the two approaches may also emerge thanks to new partnerships, in particular between producer organizations and financial institutions (Wampfler and Mercoiret 2002).

Mobilizing and Diversifying Financing Sources

Mobilization of Savings. As noted by the *Agence française de développement* (French Development Agency, AFD),

> Savings collection in rural areas generally involves low amounts—the exception being rich, densely populated areas. It is based on term deposits with cycles that are out of sync with the need for resources for loan activities. For example, cash savings tend to be at their highest at harvest time, when there are limited possibilities for recycling savings into loans. In rural areas, savings collection is a costly service. It gives rise to a very high number of small operations, making it necessary to maintain cash balances on the spot. To make this service profitable, microfinance institutions offer products that pay little or, most often, no interest. Potential savings thus tend to be invested in opportunities that are considered to be more financially or socially beneficial (purchase of livestock, housing improvements, etc.). However, there has recently been a trend toward the development of cash savings in areas facing insecurity, but the microfinance institutions are themselves fragile in theses contexts.[1]

Furthermore, for regulatory reasons, it is often difficult to use cash savings as a source of financing for loans (box 9.3). For microfinance institutions, the challenge then becomes the ability to mobilize term deposits—in particular for financing investment credit—or to function with apex structures that can ensure equalization of the cost of funds among highly diverse zones.

Bank Refinancing and Guarantees for Microfinance Institutions. Microfinance institutions often develop through refinancing from the banks. The banks intervene with credit lines for specific medium-term products (such as leasing or storage credit), by nontargeted refinancing, and by overdraft facilities that make it possible to smooth out cash flow. In view of the requirements of the microfinance institutions and the banks' possibilities, guarantees and other risk-sharing tools prove relevant once the networks have achieved a sufficient degree of maturity, and also facilitate their gradual integration into the financial sector.

BOX 9.3

Migrant Savings

Migrant remittances have become major financial flows in rural African areas in recent years. In addition to financing household consumption and, on occasion, community projects, some analysts regard these transfers as a new path for financing agricultural and rural activities through local institutions. However, the magnitude of the flows and the savings that result may well pose problems in some networks established in zones of migration (Senegal, Mali, or the Comoros, for example), because they lead to liquidity surpluses for the institutions receiving them. Indeed, while bringing migrant remittances into the banking system plays a part in local development, it nevertheless cannot palliate the structural economic problems that are present in the rural migration regions subject to considerable development constraints and that put brakes on productive investment.

Budget Resources through State Channels or through Targeted Policies. The importance of increased public support has been reaffirmed more and more forcefully in response to the challenges of agricultural transition. For Sub-Saharan Africa, in its Maputo Declaration of 2003, the African Union set a contribution objective equivalent to 10 percent of the state budget, well beyond the 3.5 percent to which public expenditure in favor of agriculture declined in 2001–02 (FAO 2006). These financial flows must, in part, pass through the various institutions of the financial system in order to support the financing of and investment in agricultural and rural productive units. The (re)construction of more effective financial intermediation in agricultural and rural areas thus leads to a renewal of the approaches and instruments related to public financing policies.

On top of credit lines, guarantee funds, and refinancing, there may also be what are sometimes dubbed "intelligent" subsidies, specifically oriented toward results and beneficiaries, to support the processes of innovation and change in microfinance institutions and enable them to improve their agricultural and rural outreach. This kind of program, which may support innovation or provide incentives to financial institutions to extend their rural outreach, is customarily conceived as temporary support, carried out in "project" form by autonomous financial institutions. Other programs are aimed at making the livelihood systems of the poorest households more secure through improved access to financial services by means of a support mechanism (such as vocational training) financed by various funds (social safety net, food security fund).

It may also take the form of subsidies (Doligez and Dufumier 2007) intended for productive investment (examples include purchase of improved dairy breeds in Madagascar, carts for transporting forage and manure in cotton areas, and incentives to purchase fertilizer for corn in Malawi) (CGAP 2006).

Financing from Commercial Banks and the Private Sector. Overall, these mechanisms have not been tested thoroughly in West Africa, but they remain potentially important: the partnership of agroindustry and producers, and the integration of commercial banks into the development of certain subsector segments, constitute avenues for further investigation.

What Conditions Should Be Implemented to Support the Financing of Agriculture?

Since the paradigm shift that, with financial deregulation and the disengagement of the state, accompanied the rapid growth of microfinance, the role of the state has changed considerably. It has gradually been reinstated in its role as overseer of currency and financial institutions (legal and regulatory framework, audit, and supervision).

The question that arises in terms of regulation of rural and agricultural finance is to what extent it needs to be specific in order to improve the supply of financial services; to facilitate the dissemination of innovations such as leasing by securing the forms of guarantees and simplifying administrative procedures; to better adapt supervisory practices to the challenges of rural and agricultural financing (by reducing, for example, the constraints of reserves and qualification standards and evaluation of agricultural portfolios) without endangering the stability of the financial system; and to create institutional spaces that foster diversity in the types of financial institutions devoted to rural and agricultural finance.

Today, in the name of "financial inclusion" (United Nations 2006), public authorities may also be prompted to implement programs intended to compensate for the market failures by providing incentives to private stakeholders, in particular microfinance institutions or stakeholders in the subsectors, to cover new demands (isolated rural areas, investments in family farming, and the like).

However, beyond this, in some countries redistributive policies are also emerging with a view to reducing, in the name of equity, the inequalities associated with access to financial services or to supporting certain sectoral development priorities, in particular agricultural ones. In some emerging countries, subsidies might be reinstated for the sake of more equitable sustainable development in order to pave the way for new specialized financial intermediaries and to reduce the costs of access to credit for the most isolated rural areas or for family producers who are excluded from the banking sector. Discounts and rebates for agricultural investment are reemerging among agricultural policy tools; in some developing countries, it is proposed that international funds be mobilized out of official development assistance (debt conversion).

Pragmatism and the search for efficiency should therefore now be of primary concern in the formulation of agricultural policies and the conditions for their implementation. It is no longer a time for formulating general and normative policy. There are incentives to devise and introduce financing instruments and mechanisms whose effects will be combined and gradually provide a "global" response to financing needs. This process should be the subject of renewed dialogue within the relevant political associations, and professional agricultural organizations should make a contribution through public policies that are genuinely coordinated among stakeholders (Gentil and Losch 2002). In this regard, the outline below lists six work areas, in which efforts can be carried out frontally with varying degrees of intensity and pace depending on the country.

- Consolidate and enhance the autonomy of the financing mechanisms and instruments that have been put to the test over the past 20 years, by guaranteeing that they have a stabilized regulatory and prudential framework, and by protecting them from any temptation to use them as deliberate instruments of agricultural policy. This applies to
  - microfinance institutions: this involves affirming their sustainability, their mission to diversify their financial and banking products fully independently from agricultural policies.
  - subsector contracts, which should accord more importance to the relationships between farmers and industry than in the past, and in which the state does not act as the stakeholder responsible for carrying out everything.
- Integrate the financing of subnational governments through decentralization policies into the strategy for financing the economic and social environment of family farms. These instruments and mechanisms may, in fact, contribute in the long term to guaranteeing regular flows of funds to finance basic infrastructure. This will create an incentivizing local environment that ensures farmers will be provided with the basic services (education, health, and so forth), which constitute a major demand of rural dwellers. This absence of services is very often cited as one of the reasons for discouragement and departure, particularly as far as young populations are concerned.
- Continue or revive innovation efforts aimed at proposing new and incentivizing financing. Two groups of farms need to be distinguished at this level. For farms threatened with marginalization, the focus should be on offering mechanisms whereby they can diversify their economic activities

and be integrated into the market. For those that can access agricultural modernization, such as intensification or diversification of production, there are a number of avenues available, such as leasing for some investments, the mobilization of farm savings in a medium-term perspective (through mechanisms of the savings-investment type), and new insurance mechanisms.

- Take up again the issue of partnerships within subsectors between agroindustries, distribution channels, and farms. Even if experience in Africa over the past decade seems inconclusive, this approach deserves to be put back on the agenda. It certainly remains one of the best ways to enable farms to obtain short-term crop financing, to ensure them access to technological innovations, and to meet the demands and respond to market signals (in terms of research and technical innovation), and it can offer a new area of actions and partnerships for producer organizations.
- Step up states' budget efforts in favor of the formation of fixed capital for farms, either productive capital, such as water works and planting perennial crops, or in the areas of research and knowledge. Change must come from the state's desire to gain maximum leverage from financial resources, know-how, and contributions of new technologies. This desire can be expressed in two ways: through choices in favor of concentration, or through long-term efforts that far exceed the usual length of projects.
- Finally, it bears recalling that questions about land tenure and rural demographics addressed in other chapters of this volume remain key. Currently these variables are often the source of decapitalization and hence constitute constraints on farm financing. Even if the formulation of short-term measures is out of the question, this issue should be a priority, because major innovations in terms of financing and investment will be possible only within the framework of structural changes in these key variables.

In conclusion, the question of agricultural and rural financing is difficult to address outside its economic, social, or political context. In view of its interconnection with numerous other issues—growth, poverty reduction, equity, territorial development, sustainable development—the approach is complicated and multifaceted, which necessitates a renewed and overarching vision of stakeholder strategies and of the role that the institutional framework can play. Nevertheless, it in no way constitutes a "passive" outcome, and a good many innovations remain to be developed in order to increase and enhance the security of agricultural investment. Moreover, with respect to access to financing, the equity objective ("pro-poor agricultural transition") must not translate into the increased fragility of an institutional fabric that, as has been observed, requires time to be consolidated. Accordingly, behind many safety nets or social compensation policies, as in the case of the new instruments of redistributive policies, the short-term approaches driven by political issues and, sometimes, certain clientelist temptations may rise back to the surface at any time. Only approaches negotiated on the basis of public policies built by and shared with all the financial stakeholders as professionals can ensure the consistency of this threefold challenge: consolidating efficient rural financial intermediation, increasing agricultural and rural investment, and combating the inequalities and social marginalization of a growing proportion of African small farmers.

Note

1. Paper for the colloquium of the FARM Foundation, "*Quelle microfinance pour l'agriculture des pays en développement?*" Paris, December 2007.

Bibliography

Autrepart. 2007. "Risques et microfinance" series 44.

Azokli, R., and W. Adjibi. 2007. *Microfinance au Bénin: évolutions et perspectives*. African Performance Evaluation Forum.

CGAP (Consultative Group to Assist the Poorest). 2006. *Graduating the Poorest into Microfinance: Linking Safety Nets and Financial Services*. Focus Note 32, Washington, DC.

CIRAD-CERISE. 2002. *La microfinance au service de l'agriculture familial*. MAE, Paris.

Consortium Alafia. 2007. *Microcrédit aux plus pauvres et distorsions de marché au Bénin*. Cotonou.

Doligez, F., and M. Dufumier. 2007. "Trajectoires des systèmes de production agricole et diversification des modes de financement des exploitations familiales dans les zones cotonnières ouest-africaines: le cas du Sud Mali." Paper for the International Conference on Rural Financial Research, IFAD-FAO-Ford Foundation. Rome (March).

FAO (Food and Agriculture Organization). 2006. *Food Security and Agricultural Development in Sub-Saharan Africa: Building a Case for More Public Support*. Rome.

FAO-GTZ (Deutsche Gesellschaft für Technische Zusammenarbeit). 2000 to 2005. *Agricultural Finance Revisited*. FAO, Rome.

Gentil, D., and Y. Fournier. 1993. *Les paysans peuvent-ils devenir banquiers? Epargne et crédit en Afrique*. Syros, Paris.

Gentil, D., and B. Losch. 2002. "Politiques de microfinance et politiques agricoles: synergies et divergences." Paper at the seminar *Le financement de l'agriculture familiale dans le contexte de la libéralisation. Quelle contribution de la microfinance*. CIRAD-CERISE, Dakar.

Joseph, A. 2000. *Le rationnement du crédit dans les pays en développement. Le cas du Cameroun et de Madagascar*. L'Harmattan, Paris.

Le Breton, P. 1989. "Les banques agricoles en Afrique de l'Ouest." *Notes et Études* 24.

Lelart, M. 2005. *De la finance informelle à la microfinance*. AUF, Paris.

Lhériau, L. 2005. *Précis de réglementation de la microfinance*. Notes et Documents 20, AFD, Paris.

Morvant-Roux, S., and J.-M. Servet. 2007. "De l'exclusion financière à l'inclusion par la microfinance." *Horizons bancaires* 334.

Réseau français de la microfinance. 2007. "Evolutions récentes dans l'offre et les stratégies de financement du secteur rural."

Revue Tiers Monde. 1996. "Le financement décentralisé, pratiques et theories," 145.

———. 2002. "Microfinance: petites sommes, grands effets?" 172.

Servet, J.-M. 2006. *Banquiers aux pieds nus, la microfinance.* Odile Jacob, Paris.

United Nations. 2006. *Building Inclusive Financial Sectors for Development.* New York.

Yunus, M. 1997. *Vers un monde sans pauvreté.* JC Lattès, Paris.

Wampfler, B., and M.-R. Mercoiret. 2002. "Microfinance, organisations paysannes: quel partage des rôles, quels partenariats dans un contexte de libéralisation?" Topic summary, international seminar titled *Le financement de l'agriculture familiale dans le contexte de libéralisation, quelle contribution de la microfinance?* CIRAD-CERISE, Dakar, January 21–24.

World Bank. 1989. *World Development Report 1989: Financial Systems and Development.* Washington DC.

Chapter 10

Building Human Capital and Promoting Farmers and Their Organizations

Jean-Claude Devèze

The human dimensions of agricultural development are too often neglected, making it particularly important to promote the professional skills of farmers and strengthen community dynamics, especially through professional organizations and by building more balanced relations among participants. Particular attention must be paid to the training of farmers and the integration of young people.

The future of African farms depends in large measure on the changes that African farmers can achieve, which makes establishing favorable conditions particularly important: secure land tenure, access to solvent markets, opportunities for developing and sharing innovations, and appropriate financing. However, it is the mobilization of human and social capital, along with personal and collective change, that, to our minds, constitutes the main prerequisite for achieving essential changes in agriculture, on the assumption that a certain number of economic and political conditions are present.

The farmer above all works individually on his farm, along with his family and possibly some "paid" workers, making it particularly important to promote individual skills and opportunities for personal growth through apprenticeships, training, information-sharing, contacts, and experience. The individual promotion of men and women and the promotion of the community are closely linked, which means that attention must be paid to the involvement of family groups; to the work in teams or in groups during training sessions; and to the way in which farmers organize themselves, increase their economic influence, enter into business dealings, and establish a vision of the missions and functions of agriculture and the nature of their profession. It is not just the farmers' human capital that must be taken into account but the human capital of the entire agricultural sector and rural world.

Behind these individual and collective transformations and these groups of participants lies the whole issue of changing mentalities, with all of its cultural, social, economic, and political dimensions. It is important that African decision makers (see part 3 of this volume) explain how they see the future changes and cultural development of their societies. This chapter merely provides some thoughts and suggests some priority work focuses to take better account of the human dimensions of the development of African agriculture and to strengthen human capital in the broadest sense of the term, that is, including social and cultural capital. Given the complexity and scope of this topic, the focus is primarily on building farmers' capacities, strengthening community dynamics, and building balanced and constructive relations among participants.

Too Little Attention Is Paid to the Human Dimensions of Agricultural Development

The chapter focuses on the following topics: the importance of sociocultural factors, the enormous lag in the education and agricultural training of rural inhabitants, and the emergence of independent socioprofessional organizations.

Among the sociocultural factors that make change difficult in agricultural societies, the following seem to be most important:

- The weight of tradition and past experience that enabled previous generations to succeed makes all outside innovations suspect, especially because the changes resulting from these innovations have been judged negatively; this can lead to the individual who makes changes without the agreement of the majority in the village being labeled misguided or irresponsible.
- With the predominance of poverty and the many uncertainties, change is synonymous with enormous risk (for example, the monetary risk for the farmer who purchases agricultural inputs or equipment that he must then reimburse). Because risk must be minimized, financial commitments are made only when the anticipated gain is substantially greater than the risk involved. The result is often a defensive attitude on the part of farmers, who focus on the short term (Devèze 1996).
- Young people very often find themselves still dependent on their parents on the family farm, leading to their initiatives being blocked. If young people set up separately, they suffer not only from a lack of resources for making changes that require operational margins, but also from increased risk in case of illness, drought, or other problems. It is therefore very important to understand the change in intrafamily and intergenerational relations (Chauveau 2005).

The social capital of traditional African societies, which has enabled them, in their own way, to establish strong relationships at the local level, is challenged by the many changes resulting from greater openness to the rest of the world and the emergence of new actors.

Given this complex sociocultural context, the changes in family units involved in agricultural and rural activities pose many problems. The modernization of the family unit is particularly difficult in that the economic environment is not secure. The human capacities of farmers, and therefore their education, must be the focus if their activities are to be transformed into family farms that can produce and sell more and better in an economic environment that they will have helped to make more secure with the support of a government that takes their problems into account.

Education and training remain major concerns as the following findings indicate:

- Families and villages play an important role in basic education, helping children determine where they fit in the way of life surrounding them and integrating the experience of previous generations; this is partly challenged by confrontation with a modern world that forces them to change by adopting new references to adjust to a broader and more complex world.
- Efforts in the areas of literacy training and education have been insufficient to reach the majority of adults in rural areas (literacy rates are below 50 percent in the countries of West Africa, with the exception of Nigeria; these rates often better in East Africa and especially Southern Africa). Current efforts in primary schools are not entirely making up the lag, especially for girls, and basic education at the primary school level remains mediocre.
- The agricultural educational system, from field training to higher education, is deteriorating almost across the board.[1] Agricultural vocational training is neglected by African governments, with most of the rare training systems in place having insufficient intake capacity and often using inappropriate teaching methods and curricula. The administration of training usually proves incapable of adapting to conditions in the agricultural sector and the specific characteristics of young people from the rural world.

Special attention must be focused on the future of young people from farming families, both male and female, who show, by their increased migration to the cities, that the condition of farmers and the image of farming as a profession are problematic (Devèze 2007). It will not be sufficient to educate and train young people to encourage them to become farmers. The conditions to persuade them to go into agriculture must be present, the problems facing all farmers must be resolved, and high-quality social and education services must be put in place.

To our mind, the most interesting advances are those relating to the growing mobilization of farmers through professional farmers' organizations (FOs). Between 1982 and 2002, the percentage of villages with farmers' organizations is estimated to have increased from 8 percent to 65 percent in Senegal and from 21 percent to 91 percent in Burkina Faso (World Bank 2008). Self-promotion of these organizations encourages strong ownership by farmers, explaining the important role played by the professional leaders who face many difficulties in establishing relationships of trust with the grassroots communities.

Little by little the other countries and regions of Sub-Saharan Africa are following at their own pace. This contributes to the involvement of agricultural leaders in rewriting public policy (Mercoiret 2006) and in international negotiations.

Five assessment criteria are proposed to estimate the strength of Sub-Saharan African farmers' movements (Gentil and Mercoiret 1991): intellectual and financial autonomy; conscious and explicit objectives; significant relations with the government or the rest of civil society; considerable political and economic influence and size, and an established internal organizational structure. Although most observers have assessed the structuring work already accomplished in the agricultural sector positively, they also believe that much remains to be done to increase the influence of FOs and national "platforms" to make them more autonomous financially and improve their operation.

To strengthen the human dimension of development, in line with the institutional changes corresponding to the priorities selected, the following three complementary approaches are explored:

- enhancing the capacities of farmers and providing vocational training of young people;
- strengthening community dynamics, particularly through professional farmers organizations; and
- building more balance and constructive relations among participants.

Capacity Building for Farmers

Innovative apprenticeships and training approaches for farmers already exist in the field, including, for example, functional literacy training centers, alternating classroom education and training in rural family homes, centers for rural professions in Côte d'Ivoire, Songhai centers providing a link between production-processing-marketing and the establishment of farmers, training of farmers by Agrisud in periurban areas in Angola, the Republic of Congo, and Gabon, agricultural training centers for young couples in northern Cameroon, and agricultural colleges promoted by the heads of agricultural organizations in Madagascar (box 10.1).

BOX 10.1

Malagasy Agricultural Colleges

In 2002, FIFATA, a federation of Malagasy agricultural organizations, decided to make initial agricultural training a priority activity. It created three agricultural colleges, which accept 118 students between the ages of 14 and 18 who have completed primary school and wish to become farmers. The aim is to provide three years of technical, economic, and social training, without taking them far from their village contexts and their parents' farms.

The colleges are managed by associations of representatives of parents, FIFATA, and local leaders.

The ultimate objective is to create a network of colleges to train young people who are then able to establish themselves professionally.

Source: Grain de sel 8.

Among the key elements needed to train young farmers capable of successfully completing their professional training and returning to their own milieu, we would emphasize the following: literacy; the interest of parents and young people in training that will prepare them to be farmers; the willingness of parents to pass their knowledge on to young people in training sessions and to open up to new knowledge; the availability of trainers who are able to adapt the training to the needs of young people and the family context; and recognition, by the government, donors, and territorial authorities, of the relevance of innovative training methods and an interest in supporting them financially on a sustainable basis.

Once trained, young farmers often face numerous difficulties in their chosen profession. The problems faced by a young Burkinabè farmer returning to the family farm are one example: he suggests to his father that they employ draft animals, using money deposited in an account with a savings and loan association, but his father says that he must first come to the aid of a family member in difficulty. Once this obstacle is overcome with the help of mediation by a local leader, the young farmer finds it difficult to break in his cattle and they escape into the bush. When they are finally rounded up, he is gradually able to control them with the help of practical advice and thus to demonstrate the usefulness of draft animals for the family farm. This shows the importance of a capacity to make decisions while maintaining positive relations with the head of the household, the roles that can be played by experienced advisers or leaders capable of proposing constructive compromises, and the need to learn techniques and to adapt them to the local context.

Particular attention must be paid to training women, who represent half of the agricultural workforce and who play a major role in domestic transport, marketing and processing of agricultural products, and other income-generating activities. A change in how farm families function and the development of women's creative and relational capacities must be explored further.

Advisory services for family farms, based on learning processes that combine technical, management, economic, and financial techniques, represent an original approach to promoting decision making by farmers and their families (box 10.2). This updated approach to agricultural extension—begun in West Africa in the context of projects supported by Coopération française (Gafsi et al. 2007)—is being adopted by more and more farmers' organizations to support their members.[2]

The many initiatives in the areas of training, apprenticeships, and advisory services must be supported, but they are not sufficient to the task at hand. The training and integration of a new generation of enterprising young farmers and agricultural and para-agricultural specialists capable of working with them represents an enormous challenge that is related to the challenge of integrating large cohorts of young job seekers into the labor force. Meeting

BOX 10.2

Family Farm Advisory Services

During a workshop in Bohicon, Benin in 2001, an ambitious approach to providing advisory services to family farms—covering everything from management of activities (for example, a family granary) to advisory services for the entire farm—was developed on the basis of numerous experiments carried out in West and Central Africa in the context of projects aimed at improving farming systems.

This approach was based on building the capacities of farmers in a group learning situation (for example, a couple dozen volunteers consisting of farmers and farm managers or literate young people in the confidence of the head of the farm). Data tracking and recording tools made it possible to measure, analyze, and make decisions. This innovative approach had various results in several areas, including more intensive farming methods, organization of work on the farm, methods of ensuring food security, management of monetary flows, and methods of preserving soil fertility. Family discussions on the quantitative results obtained also enabled them to make decisions on cash flow, remuneration, or investments.

There are also many indirect effects, such as increased demand for literacy training, a ripple effect locally through increased technical exchanges, greater demand from farmers for services from their farmers' organizations, and higher-quality discussions on decisions to be made.

The arrangements involve the use of top-notch specialists, who must be trained and remunerated, creating a need to cover costs through subsidies as well as through the financial participation of the beneficiaries.

this challenge requires "conceptual tools to convince decision-makers to implement national programs to reforge agricultural and rural training" (Debouvry 2007).

The process for reforging agricultural training (combining vocational training, on-site apprenticeships, and continuing education) must be based on the fundamental principles of equity (the right of all participants in the agricultural sector to obtain training), internal efficiency (ensuring that all students obtain training in the minimum possible time), external efficiency (relationship between training and employment to avoid training students who are not suited to the market when they graduate from costly programs), and effectiveness (cost-benefit analyses).

To meet this challenge, it has been proposed that African expertise be quickly developed to take responsibility for promoting and working on this issue, that awareness-raising be stepped up among the various groups of national and international actors concerned with this issue, and that agricultural training strategies be included in the broader context of a rural and agricultural policy favoring the integration and establishment of young graduates (Debouvry 2007).

The conference on vocational training recently held by the United Nations Educational, Scientific, and Cultural Organization considered the factors for making vocational training attractive to persons from the informal sector (Walther and Filipiak 2007). It emphasized the need for recognition of the sectors concerned as fully professionalized, a strong focus on future jobs, the development of the role of professional organizations, and synergies between generations. A current debate is focusing on the usefulness of agricultural vocational training in contributing to the transition from the still dominant informal sector to the formal sector as part of the agricultural transitions requiring a change from family units of activity to business-oriented farms. To our minds, the important issue is more that of greater autonomy for farmers and their organizations in their choices, which can promote the transition to a fiscal policy that will enable the agricultural sector to secure greater financial resources to address its priorities.

Promoters of vocational agricultural training (Debouvry 2007; Maragnani 2007) rightly insist on the need to reach a critical mass of young farmers. The training-apprenticeship-counseling processes are a priority for the future of young people. The difficulty lies in convincing decision makers of this and in guiding the choices made, given the costs and constraints to obtaining high-quality training and bearing in mind successful experiments to use as examples. Most countries have to overhaul their agricultural educational systems in coordination with the trainees, their relatives, promoters of initiatives, and farm organizations, which explains the importance of finding synergies between the public authorities and nonstate actors. At the same time, thought must be given to the training of those who wish to leave agriculture to enter other sectors.

In rural areas the education and training system faces three challenges: promoting high-quality universal education; reforging vocational training that favors the training of young people in agricultural, para-agricultural, or other professions; and supplementing vocational training and education with on-the-job apprenticeships and continuing education. To that end, education, vocational training, apprenticeships, and continuing education must be seen as a social enterprise involving various participants in rural areas around shared objectives. The capacity of farmers to change depends not only on training and apprenticeships such as those proposed above, but also on all the other factors that give meaning to their personal transformations and help them feel sufficiently free to undertake and participate in collective initiatives.

Strengthening Collective Dynamics

Promotion of the individual is closely linked with promotion of the community and therefore with the way in which the social dimension is taken into account, starting with vocational training or group learning from family farm advisory services. Collective capacities can be mutually strengthened, whether within a family, within in a training or advisory group, or within a professional farming organization.

Within families, good relations among family members can promote both the fulfillment of personal dreams and the implementation of group projects. Take the example of an extended family with an acknowledged head of household who has been able to develop the family farm by using draft animals, diversifying the farm's sources of income, expanding acreages under crop, purchasing cattle, and placing his savings with the local credit union. The eldest son has taken literacy courses and received training from the agricultural advisory services to assist with the technical and economic choices on the farm and family decisions on remuneration and investment. The time will then come, for example, to purchase a tractor, which will be assigned to a young family member who is interested in mechanics, or to consider the option of raising rabbits, which interests a female family member, and so forth. In the field, there are many examples of successful family advancements such as these, but also many cases with much less favorable outcomes: the migration of the men, dissension with the father or among family members, the woman left behind with the children, etc. Often, too little attention is paid to the family dimensions of change on the farm and the role of women in these processes (Wilhelm and Ravelomanantoa 2005).

The progress made by agricultural professional leaders and their organizations, despite the obstacles that appear to face them, should not cause us to overlook the difficulties encountered within FOs:

- These young organizations often lack administrative and management capacities, which detracts from their actions.
- Some leaders are overburdened, preventing them from taking time to reflect on their organizations' strategies, train the next generation, and work with the grassroots members, administrators, and staff of their organization.
- The staff of FOs may be tempted to usurp some of the power.
- The "turf protection" culture shown by too many male politicians seems to rub off on some agricultural leaders, who work harder to maintain their positions and the related benefits than to perform their mandates.
- Schisms[3] at the regional and national levels tend to recur repeatedly, seeming to be fostered in some cases by political leaders or encouraged by irresponsible external financing.

All this explains the reactions in the field of members of FOs, who complain about the such issues as the poor functioning of their groups, the compromises their leaders make with the authorities, and the lack of transparency in the accounts. This is why it is important that FOs continuously improve their democratic governance, enforce the by-laws on term limits for leaders, promote management boards that provide transparency, control the use of external financing, and combat dependent relationships between some leaders and the authorities.

One focus for FOs is certainly ongoing training for their leaders and members, as well as for the experts they employ. The difficulty lies in organizing training that is consistent with the FOs' strategies, while ensuring that the stated objectives are shared by all. Other problems include finding financing, selecting trainers who can adjust their approaches to those being trained, and assessing the results. Synergies must also be found with the systems for the agricultural vocational training of young people.

Advisory services to FOs should be a new priority to help them improve their governance and therefore their credibility. The introduction of this type of arrangement in the cotton-growing areas of Burkina Faso and Mali and in irrigated zones under the authority of the Office du Niger in Mali and in the Senegal River valley (box 10.3) has had significant results in improving the transparency of accounts, combating FO indebtedness, and clarifying decisions to be made. The entire issue now is sustaining the arrangements already in place, helping them evolve to better respond to needs, financing them, and expanding them in the context of strategies to be defined with FO leaders.

BOX 10.3

Farmers' Organization Management Centers in Senegal

Two rural economy management centers (CGERs) were created in 2003 in the Senegal River valley and delta to provide farmers' organizations (FOs) with an accounting and management mechanism managed by their leaders. These centers are supported by a coordination center that provides methodology, trains management advisers and leaders of FOs, and helps them use the results to assist with their decision making.

Training has been provided for 3,500 leaders in the structuring and operation of FOs, the day-to-day management of an organization, planning, understanding financial statements, managing credit, and the like.

In this way, 200 FOs have been provided with management tools, internal procedures, and properly prepared accounts, and 160 presented their financial statements in annual meetings in 2006.

The FOs provide only 40 percent of the financing of the CGER, explaining the need for additional financing from the partners, particularly the Senegal River Valley and Delta Development Company and an AFD project.

These training and management efforts should gradually lead to the democratization of the FOs, the replacement of leaders with a new generation, and effective internal controls based, for example, on the creation of oversight committees. Collective change concerns not only the FOs but also other informal and formal institutions in which farmers are involved, including village committees, microfinance institutions, local communities, and training centers. To better understand the power relations, dynamics, and constraints, it is essential to analyze the interplay between the actors involved.

Construction of Balanced, Dynamic Relations among Actors

Here the focus is on the relations between farmers and other sector participants. Farmers are too often considered the "weak link" in interprofessional relations, explaining the need for a more objective analysis of the responsibilities of each party in the face of the difficulties in building the future together.

Merchants and buyers of farm products have too often profited from their larger operational margins, a situation that entices farmers to use various tactics if the opportunity arises, such as selling to buyers other than the one that advanced them funds for their crop or trying to cheat on the quality. In organized subsectors, such as the cotton subsector, buyers and ginning companies have often conducted business in complicity with local authorities, making it difficult for cotton producers to assert themselves as full-fledged partners.

Building farmers' capacities, in some cases with the help of the return to the farm of educated persons or persons who have gained some experience during migrations, should help them gain respect and enable them to make proposals from positions of greater strength.

Relations with governments have often deteriorated owing to the disdain officials show for farmers, the failure to take the needs of the rural population into account, and the failure to respect the rule of law. Farmers often consider government officials to be the "white man's lackeys" and refer to them as "government flunkies" in their local languages.

Nongovernmental organizations and service providers are essential intermediaries to help smooth relations with farmers and implement project actions. Acting as "development brokers" (Bierschenk, Chauveau, and Olivier de Sardan 2000), they can play a useful role, but there is a risk that they will defend their own interests more than the interests of the farmers.

Relations with donors are often contaminated by the financial dependence of rural populations, as reflected in the long lists of "grievances" raised at some village meetings. This problem particularly arises in major centers, where leaders spend a good part of their time trying to mobilize external financing. Moreover, unkept promises and false hopes make villagers more mistrustful. Finally, the increase in the number of projects and actions without effective coordination in the field does not help farmers control recourse to external assistance. All of this often makes it difficult to construct relations of confidence with farmers or national political leaders, which are essential for identifying common objectives based on mutual interests in the long term.

The reactions of the people to development projects or other external interventions have been analyzed at length by sociologists and socioanthropologists, who study development institutions, governments, and the populations they address, as well as the interactions between "the developers and the developed" and the strategies of those involved, who come from different social contexts and are brought together by development policies and practices (Olivier de Sardan 2007). This type of research is rare in the field of agriculture, however. If it exists at all, it is often obscure or not fully taken into account by local leaders or external partners.

It is particularly interesting to analyze the relations among those involved in the development of national agricultural policies. An important advance, as shown recently in Mali and Senegal, has been the involvement of FO leaders in the development of laws defining agricultural policies. In the case of Senegal, implementing mechanisms are lacking for the 2004 Framework Law on Agriculture, Forestry and Livestock Raising. The ability of the National Council for Rural Cooperation and Dialogue to intervene positively in a top-down initiative by the president of Senegal is particularly important. However, the lack of "clarification of objectives, agricultural development strategies, operational programs, and appropriate budgets" (Niang 2007) is criticized. One question is whether the various parties involved in the development of agricultural policies have understood all the implications of the various views on the future of the rural societies they wish to promote, and whether the compromises in the framework law between those who defend the family farm and those who promote agribusinesses are based on a sufficiently shared intent to achieve a policy that can be applied. The "great agricultural offensive for [food] and abundance" launched by the president of Senegal without prior discussions in the spring of 2008 does not augur well in this regard.

The development of regional agricultural policies faces similar problems, with the additional issue of the influence of large international donors, as reflected in the difficulties encountered in negotiating economic partnership agreements with the European Union (*Grain de sel 2007*).

Another way of looking at the relations between participants is to analyze how institutions function and how they have evolved. The case of interprofessional sectoral organizations (see Inter-réseaux Développement rural's working group) provides many lessons on how African participants are taking ownership of new organizational concepts. It is often difficult for them to define realistic priority objectives, to limit the partners involved to the core group of farmers and their buyers, and to clearly establish the role of the government.

It is also important to examine how the role of the government can be adapted to the changing paradigm, which means moving from prescriptive policies that favor government entities and the supervision of farmers, to policies that are mutually agreed upon and focus on guiding the efforts of participants to develop their own sector.

An important way of looking at the relations among participants is to analyze power relationships and the way in which the participants use these relationships. There is no doubt that things have changed with the emergence of civil society, particularly the FOs. The strikes in the cotton delivery subsector in Mali and Côte d'Ivoire have had a varying degree of impact, depending on circumstances, the quality of the FO leaders, and the degree of cooperation among them. These strikes also raised questions about the effectiveness of the methods to follow, such as, "not delivering the cotton crop" or "not sowing the cotton crop." The actions of national organizations—with the support of international NGOs—to challenge the importation of frozen chickens or powdered milk competing with small-scale local production have had positive short-term results in general. But the question remains of improving productivity to enable farmers to compete with more competitive intensive farming methods, both nationally and internationally. At the international level, the countries of the South have been able, with the help of their agricultural leaders, to influence

WTO negotiations by raising the issue of the future of African cotton, which is endangered by subsidies for farmers in rich countries, but no tangible results have yet been achieved (see chapter 8).

In these power relationships, the playing field is not level among all participants, owing to inequalities and asymmetries, which raises the difficult issue of how to promote equity by combating inequality. Using an institutional analysis that looks at opportunities for change, the following approaches can be promoted: resisting inequalities and dominant relationships, mobilizing in the long term enlightened forces (elites, experts, external supporters) to respond to the aspiration toward greater equality and to establish participatory approaches, and constructing networks for making political decisions and leading institutional change (Bebbington et al. 2007). Strengthening professional farming organizations should enable them to build balanced alliances with the government and constructive contractual relationships with their main partners.

Conclusion

The attention focused on the problems of feeding our planet should not make national and international leaders overlook the fact that the solution to these problems lies, first of all, in promoting the agricultural sector. While it is important that the sector be provided with a favorable economic environment, it is essential to develop its human capital: not just farmers, but government officials, experts, and researchers, contractual partners in the production sectors or in projects, etc. This requires considerable political will to channel momentum toward joint projects, such as training for young farmers to ensure their success on their farms, time (20–30 years) and continuity of action to ensure progress with the training and increased responsibility of these young farmers, and targeted financing in the long term with the help of the parties concerned.

To mobilize energies toward strengthening human capital in the agricultural sector, the first priority is public policy, followed by the capacity to make credible proposals based on the beginnings of solutions already found in the field, and finally by the ability to mobilize human and financial resources to put such solutions into action. Until professional and political leaders become aware of the urgent need to act, it remains possible in the meantime to strengthen and disseminate worthy initiatives.

Notes

1. The disappearance of applied engineering training at the Institut agricole de Bouaké in Côte d'Ivoire, the inability to promote viable centers for interstate higher training in francophone Africa, and the deterioration of education in agricultural *lycées* in Madagascar are some examples.

2. Inter-réseaux Développement rural (http://www.inter-reseaux.org).
3. Some examples are the difficulties encountered with the establishment of the African Cotton Producers Association (APROCA) following the refusal of the West African Network of Farmer and Producer Organizations (ROPPA) to create a specialized cotton commission, the many organizations created in Senegal recently, and the splits in the cotton-growing organizations in Benin.

Bibliography

Bebbington, A., A. Dani, A. de Haan, and M. Walton. 2007. *Institutional Pathways to Equity: Addressing Inequality Traps.* World Bank, Washington, DC.

Bierschenk, T., J.-P. Chauveau, and J.-P. Olivier de Sardan, eds. 2000. *Courtiers en développement: Les villages africains en quête de projets.* Karthala, Paris.

Chauveau, J.-P. 2005. "Les jeunes ruraux à la croisée des chemins." *Afrique contemporaine* 214.

Debouvry, P. 2007. "La formation de masse face aux enjeux de développement des exploitations familiales rurales ouest-africaines." Contribution to the Réseau FAR seminar in St. Louis on training and advisory services for the promotion of rural family farms).

Devèze, J.-C. 1996. *Le reveil des campagnes africaines.* Karthala, Paris.

———. 2007. "Les jeunes, bâtisseurs de l'agriculture de demain." *Grain de sel* 38.

Gafsi, M., P. Dugué, J.-Y. Jamin, and J. Brossier. 2007. *Exploitations agricoles familiales en Afrique de l'Ouest et du Centre.* Quae, CTA, Paris.

Gentil, D., and M.-R. Mercoiret. 1991. "Y a-t-il un mouvement paysan en Afrique noire?" *Revue Tiers Monde* 128.

Grain de sel. 2007. "Accords de partenariat économique: presentation, analyses, points de vue." *Grain de sel* 39.

Maragnani, A. 2007. "La problématique de la formation professionnelle dans le secteur agricole et dans le milieu rural." Presentation at the GEFOP workshop on professional training in the agricultural and rural sectors. UNESCO.

Mercoiret, M.-R. 2006. "Les organisations paysannes et les politiques agricoles." *Afrique contemporaine* 217.

Niang, T. 2007. "Le cas de la loi d'orientation agrosylvopastorale (LOASP) du Sénégal." *Modalités de dialogue entre societe civile et État.* Réseaux Impact. Paris.

Olivier de Sardan, J.-P. 2007. "Vers la socioanthropologie des espaces publics africains." *Revue Tiers Monde* 191.

Walther, R., and E. Filipiak E. 2007. *La formation professionnelle en secteur informel.* Notes et Documents 33, AFD, Paris.

Wilhelm, L., and O. Ravelomanantoa. 2005. "Première approche de la problématique famille/genre/jeunes ruraux pour appréhender le devenir des agricultures familiales autour du Lac Alaotra à Madagascar." AFD, Paris.

World Bank. 2007. *World Development Report 2008: Agriculture for Development.* World Bank, Washington, DC.

Part 3

Cross-Cutting Views

At a time when the voices heard reacting to the problems of hunger in the world are primarily those of international leaders and intellectuals from the developed world, it is important to heed what African leaders have to say about the future of their agricultural systems. Considering that the statements made by altogether too many African heads of state are wanting in consistency and that not enough thought is given to this area by the majority of urban elites, it is important to give a voice to three leaders of regional small farmers' organizations and to one former political leader who has been active in promoting family agriculture.

From their thoughts emerge the major components of the ongoing debates on the future of African agriculture, in particular the following:

- Most political leaders, after having overlooked the rural world for too long, are at pains to properly assess the magnitude of the challenges facing African agriculture and to take into account the differences operating in the various different types of agriculture.
- The promotion of family agricultural systems is of utmost importance for reducing the deep gaps between the city and the countryside and between the rich and the poor, and also for safeguarding values such as solidarity and sharing.
- The professional agricultural organizations have a decisive role to play alongside the national and regional authorities when it comes to defining agricultural policies.
- There is a need to combat unbridled deregulation, by controlling it according to the level of development attained and the objectives set (food sovereignty, enhanced productivity, combating inequalities).
- It is a matter of priority to make agriculture more attractive to youth, to make land tenure more secure, to promote access to financing, and to reinforce the farmers' capacities and, first and foremost, the capacities of the members and leaders of professional organizations.
- It is necessary to increase the financial resources devoted to agriculture, and most important, the corresponding state budgets, but efforts must also be made to interconnect national and regional public policies in the context of international negotiations.
- Donors, which must forswear the desire to impose their views, must support an enhanced commitment on the part of African states and regional organizations in favor of the agricultural sector.

An analysis of how agricultural leaders each approach the various difficult issues listed below reveals a number of points that would merit further discussion between them:

- The problem of the sizable inequalities among family farmers is more often than not disregarded by leaders, who prefer to take a global approach to their defense of family agricultural systems.

- "Population growth, by itself, can have catastrophic consequences," compared with "It's not population that causes a problem; the problem is insufficient production and the poor economic conditions in which people live."
- "Agribusiness has its place, especially if we have sound infrastructure; if agroindustrial enterprises do not stay just in the city they can capture some of the migration toward the cities and attract young people seeking opportunities in the agricultural sector," compared with "We have nothing against an agribusiness that invests on a clear basis without expropriating small producers, which risks compromising the future of coming generations. It can expand in zones where there is still agricultural space awaiting improvement."
- "Family agriculture is recognized by governments and by development partners, such as the European Commission," compared with more pessimistic viewpoints on how little this silent majority has been taken into account.

Various other important questions also remain to be examined in greater depth by agricultural leaders, ranging from those relating to changes in the status of the various members of farm families (women, youth, social inferiors), to the farming structures to be promoted in order to make them more productive, to the role of professional organizations in the management of markets and agricultural subsectors, to the training of farmers.[1]

It is on the basis of this corpus of positions taken by African agricultural leaders[2] that real cooperation must be initiated or continued regarding the approach to meeting African agricultural challenges. Will African political leaders and urban elites shed their negative attitudes toward agriculture and the rural world to finally create relations of trust based on developing credible, concerted positions and implementing the ensuing actions?[3] The future of the silent agricultural revolution promised by an increasing number of smallholder leaders depends in large measure on the answer to this question. Another important factor will be the manner of promoting, on the global level, the complementarities between the agriculture of the North and the agriculture of the South, with a view to feeding the planet.[4]

Very importantly, the creation in Addis Ababa on May 23, 2008, of the Pan African Platform of Farmer Organizations constitutes a significant step in the organization of the African agricultural world. (The text of the declaration is reproduced below.)

Pan African Farmers Platform

Final Declaration

Faced with the alarming situation that has struck the African populations, the networks of farmer and agricultural producer organizations of Southern Africa (SACAU), Central Africa (PROPAC), Eastern Africa (EAFF), and West Africa (ROPPA) met in Addis Ababa, Ethiopia, from 21st to 23rd of May 2008, to share information and exchange ideas on the current state of African agriculture and possible solutions.

Considering that networks of African farmers' organizations all have the same mission, i.e. to defend and promote the interests of agricultural producers;

Noting that these African agricultural producers share the same geographical space and natural resources: land, water, forests;

Noting also that, although they represent the demographic majority of the African population, these family farm households and agricultural producers still suffer the consequences of agricultural and rural policies that do not reflect the realities they live and the preoccupations they continually proclaim;

Noting also that, thanks to the sweat of their labor, which is badly remunerated thanks to constantly decreasing agricultural prices, the States—on the contrary—have been able to harvest significant wealth which has very often been invested elsewhere than in the rural areas;

Noting finally that, today—as yesterday—these agricultural producers are the main victims of conflicts, disasters and crises such as the current one on food;

The networks of African farmer and agricultural producer organizations reviewed the different factors that are at the origin of the food and agricultural crisis in Africa.

It must be recognized that, despite efforts to promote regional integration, most of the actions and initiatives are seriously behind schedule. On the contrary, despite the aspirations of NEPAD, Africa continues to be oriented more towards the outside than inwardly.

African agriculture has thus encountered a failure in which all of us have participated: we Africans in the first instance, African political leaders and farmers' organizations, as well as our partners and the bilateral and multilateral cooperation programmes.

The farmer organization networks consider that the present situation of African agriculture is bad. However, they judge that this is not a fatality and that the situation of food price increases is not necessarily an unfavorable factor.

Seizing the current opportunity for African farmers to obtain a better remuneration for their products, however, requires that our States, our Regional Economic Communities and the AU urgently engage in a dialogue involving all of us, here in Africa and not elsewhere.

The farmer organization networks also noted that over more than five years they have strengthened their mutual knowledge and have built up a real spirit of solidarity through concerted action, in particular while working together to improve the feasibility of the NEPAD and to warn the world of the threats which the EPAs might pose for the future of African agriculture.

These challenges have convinced them that the progress of African agriculture can only be lasting if the farmers' organizations can act at continental level. The four networks of farmer organizations affirm, through this declaration, their total engagement to assume this historic necessity by deciding, here in Addis Ababa, to establish a "Pan African Platform for the farmer of Africa."

The farmer organization networks have established a steering committee composed of the 4 presidents of the 4 sub-regional farmer organization networks and have designated Mr. Mamadou Cissokho as facilitator.

This new instrument, in our eyes, brings a strong value added to the pursuit of the mandates and the activities of our local, national and sub-regional organizations. It also constitutes a powerful lever to promote a resurgence of African agriculture so that it can fulfill the functions of any agriculture worthy of its name.

Conclusion

Convinced that there are no alternatives to the mobilization of our own human resources and our own financial resources, however modest they may be, and conscious of the fact that our continent—despite the negative image of the outstretched hand, of suffering, of misery that is projected to us every day—possesses natural resources, high quality human resources, and positive values that are applicable to all of humanity, we commit ourselves, in the context of the Pan African platform of farmers organizations, to save our lives, our families, our nations and Africa, our continent.

We, the undersigned,

Mrs. Fanny Makina	Vice President	SACAU
Mr. Philip Kiriro	President	EAFF
Mrs. Elizabeth Atangana	President	PROPAC
Mr. N'Diogou Fall	Former Chairman	ROPPA

Notes

1. Jacques Faye, the Senegalese sociologist, raises questions (in the electronic forum "*Nourrir le monde*" [Feed the World] organized in Spring 2008) about the real importance of professional organizations in managing subsectors, about their capacity to defend their members and provide them with the services required, and about the influence of downstream structures on agricultural leaders, among other things.
2. See the West African Network of Farmer and Producer Organizations (ROPPA) website, for example, the "Appel des paysans et producteurs de l'Afrique de l'Ouest membres du ROPPA aux chefs d'État et aux honorables députés des parlements nationaux et du parlement de la CEDEAO" [Appeal of the West African small farmer and producer members of ROPPA to the Heads of State and honorable Deputies of the National Parliaments and WAEMU Parliament], Ouagadougou], April 30, 2008 (http://www.roppa.info).
3. The following quotation illustrates the importance of restoring relationships of trust: "The officials never respond to our queries; we don't trust them. When it comes to lying, they lead the pack" (reported by René Lefort in "La révolution verte en panne" [Breakdown of the Green Revolution], an article on Ethiopia published in *Le nouvel Observateur,* May 8, 2008.
4. "A strengthened partnership between small farmer organizations of the South and the North" can play an important role in this area, as advocated in the 2008 overview report of the French Farmers and International Development organization in Paris (http://www.afdi-opa.org).

Chapter 11

Views of a Senegalese Agricultural Leader

N'Diogou Fall*

Family agriculture, the method of enhancing value that is most appropriate to the realities and interests of the rural world, is at the core of the thinking and actions of agricultural leaders of West Africa. There is a need to remind political leaders and those responsible for official development assistance that agricultural policies are first and foremost the concern of farmers before being the concern of others, that whenever necessary family farm production must be protected from imports so that African products satisfy local demand, and that the rural citizenry have the same rights as city dwellers.

What is your vision of the future of family agriculture in Sub-Saharan Africa?[1]

For us, it is impossible to conceive of the future of agriculture without placing family agriculture at the core of our thinking. Certainly, the family farm has characteristics that are economic, social, and even sociocultural. It is the method of enhancing the value of rural resources that is the most appropriate to the realities and interests of the rural world. It has production and reproductive systems that maintain and improve the life of the group while integrating the values of solidarity and sharing. More and more often, its importance is no longer contested, and the consideration given to family farming is such that it is beginning to be taken into account at the national and subregional level, in particular in the development of agricultural policies.

---

*Former Chairman of the West African Network of Farmer and Producer Organizations (ROPPA).

Is small-scale agriculture socially, economically, and environmentally viable? Is there a way to prevent increasing inequalities among small farmers?

For us, small-scale agriculture is the best track. It is our present reality. It is the least costly approach; one that is most within reach of the support capacities currently available for agriculture.

African family farms have a central role to play in the future of African agriculture. Well buttressed and well supported, they constitute a real future solution. The family farm is nothing other than an agribusiness built around the head of household, and often consisting of 10 to 20 persons. In Africa, these farms are the most widespread agricultural system.

In Senegal there are nearly 400,000 such farms. They control 90 percent of the land and account for 90 percent of national production. In the West African Economic and Monetary Union (WAEMU) and Economic Community of West African States (ECOWAS) countries, the situation is roughly the same. That is why, at ROPPA, all agree that family farms must be protected. This is because the primary aim of our agriculture is to feed our undernourished populations. To treat the concept of family farms pejoratively is tantamount to seeking to confine this farming approach to a form of traditional agriculture that does not evolve. We have been reproached for not diversifying our agriculture. But this is a failure to understand that the African small farmer has always endeavored to minimize risks, and that this has led to some diversification. In any African family farm, the small farmer is raising chickens, goats, or sheep, as well as marginal food crops such as *niébé* or, in Senegal, *bissap*, and so on.

What is the capacity of small farmers to effect change in a rapidly evolving world?

Small family farms are more flexible and have demonstrated their capacity to adapt in the course of crises. This capacity to adapt is far superior to that of the large private and commercial agribusinesses.

At ROPPA, no one questions the benefits of modernizing agriculture. In the Senegal River Valley, which is shared by Mauritania, Mali, and Senegal, more and more family farms are using localized irrigation, tractors, combines, fertilizers, and pesticides for rice and horticulture.

Unfortunately, the free market policies of the past two decades, as well as globalization, have led to the crises experienced by African agriculture. The repercussions of opening markets, imposed on WAEMU and ECOWAS by the international organizations and the European Union, are reflected in agricultural prices for our output that have been on a downward path.

In Senegal, imports of frozen chicken, rice, and the like are destroying local subsectors. As a result, there has been a sharp drop in small farmer incomes. There are even situations in which the producer cannot even cover the costs of production in order to purchase agricultural materials and inputs. All this exacerbates poverty in rural areas. One of the consequences of the absence of protection for family farms is the breakdown of agriculture, with the resulting disparities and inequities observed in the distribution of global wealth.

Could you take stock of the implementation of the agro-sylvo-pastoral framework law (LOASP)? What kinds of implementation problems to you see?

The main impediment to implementation of the LOASP is a lack of political will. However, this law is revolutionary in the sense that it was negotiated with the stakeholders concerned, in particular the small farmer organizations. However, in its implementation we see some lack of courage to move forward in the spirit of this consensual legislation.

Is it possible to have a common project with African states and political leaders regarding African family agriculture?

ECOWAP (the common agricultural policy of ECOWAS) demonstrates that this is feasible. Today, most of those we talk to agree that African family agriculture offers more possibilities for resolving problems of a socioeconomic order, the issues of job creation and income redistribution. It has finally found its place in national and subregional projects. It is, in fact, possible to reach a consensus with regard to its importance and preponderant role in agricultural policies. When it came down to drawing up an agricultural policy for West Africa, one background question was always this: do we promote large-scale industrial agriculture, which assumes a concentration of land-holding and hence an elimination of the small family farms and consequently increasing exclusion, or instead do we place in perspective an agricultural policy based on family farms while seeking to modernize them, enhance their productivity, facilitate their access to resources, and create an environment conducive to development?

To our way of thinking, a viable agricultural policy in West Africa of necessity means choosing the second of these alternatives. And beyond the vision, there is a need to reach understanding with regard to the instruments, the tools that have to be used to channel and achieve that vision. That is what we have endeavored to do in the context of the ECOWAS agricultural policy. While we may not have won the day entirely, we are, nonetheless, pleased that this policy does stress the importance of small-scale family agriculture.

We are also pleased that this policy stresses food sovereignty, an issue that is of particular importance to us. And why? It is because the first form of sovereignty is food sovereignty. If we do not control our food supply, we are weakened and become dependent. Placing the emphasis on agricultural policies that do not enable us to be independent "as regards food" thus appears to me to be an important issue. And this is the first time in any policy for the West African region that this concern has been the subject of clear expression and some form of commitment. While this decision has yet to be reflected other than in writing, it is up to the authorities to pursue the agricultural development of the region on this basis.

However, in order to improve its support, African family agriculture would benefit from being better known, from a greater appreciation of its dynamism, potential, and constraints. In this context, research and scientific inquiries in general have a role to play. They can contribute to this appreciation of its real performance and the attendant constraints.

What is the future for young people? Should all of them stay on the land?

Properly addressing the future of young people should be assessed in a process that will take time. Currently, young people see no future in agriculture. It is necessary to revitalize the hope for an agriculture that can "feed the farmer" and ensure that all farmers live with dignity. Should this prove to be the case, young people will be interested in this line of work. Moreover, there is a need to create a "virtuous circle" that will make it possible to create other trades based on agriculture. In this case, young people who are not farmers will not be excluded from the rural world but instead will move into these new trades created thanks to agriculture. The aim is to think toward evolving to a "renewed rural economy." This means rethinking agriculture more comprehensively, because it is part and parcel of revitalizing the dynamism of the rural world through the creation of jobs and services.

Can rapid population growth be transformed into a positive development? Should an effort be made to reduce population growth in rural areas?

Population is not the problem. The problem is insufficient production and the poor economic conditions affecting the people. There are no automatic responses available, such as: "by cutting the birth rate by 50 percent we will improve living conditions." On the contrary, it needs to be borne in mind that the population makes up potential consumers and thus creates demand. For our economies, it comes down to achieving the performance needed to take advantage of this market opportunity. When your population is growing at about 3 percent and you have economic growth of 7 or 8 percent, this is not a source of concern, in particular if the benefits of growth are well distributed.

What is the proper place for agribusiness? What about relations with them?

We have nothing against an agribusiness that is investing on a clear basis so long as it is not expropriating small producer land and thereby threatening the livelihood of future generations. These undertakings can grow in areas where there is still farmable land to be developed.

If an option is taken that would favor foreign capital taking over land, it could have serious consequences. In fact, when the vast majority of the poor are deprived of their meager resources, thereby denying them access to a minimum income, this could induce social conflicts that would be impossible to control.

Moreover, if the agricultural policy choices place 70 percent of the resources into this agribusiness to the detriment of family agriculture, we do not agree and will demand an equitable economic policy. In fact, for us it is a question of asking that family agriculture be given its rights in the name of the equity that should benefit all social stakeholders.

What is the ROPPA strategy for promoting family agriculture?

We are promoting family agriculture in the context of formulating agricultural policies that we regard as essential and unavoidable. But they must be developed in a participatory manner. For us, it is no longer the time for a handful of technicians to sit around a table and draw up national or regional policies. It is important because participation makes it possible for all stakeholders to express their concerns and reach a consensus. It is only under these conditions that the policy will be appropriate and each and every stakeholder will see that his or her ideas have been taken into account. Collective commitment is necessary for the implementation of an agricultural policy. This is the case, for example, of the ECOWAS policy. All the concerns of the stakeholders were taken into account.

For us, it is impossible to consolidate regional integration and create a dynamic regional market, one that is profitable for farms and the West African private sector, without appropriate protection measures. This is why we have been fighting with all our might to influence our decision makers to increase the common external tariff (CET). Our objective is to achieve the adoption of the creation of a fifth tariff band aimed at ensuring the rebalancing of appropriate levels of protection between the region and its main competitors, and an appropriate protection for sensitive or strategic agricultural and industrial products, for the regional integration of markets, economic development, poverty reduction, and food sovereignty.

What problems does your organization face?

Our main problem in West Africa is associated with public policies. There are no coherent policies. In the final analysis, we are up against foundering systems in terms of thinking, resources, and regulation. As a result, everything that happens in our environment affects us directly. Our agricultural systems are completely exposed. The fundamental question is this: how do we manage to construct policies that correspond to our real capacities from the standpoint of finance, technology, and so on? Our political decision makers are functioning with models that were designed in particular circumstances far removed from our realities. They are only methods from elsewhere, like the one that involves "developing biofuels to combat hunger." Politicians should understand that our position is to defend family agriculture, and that it deserves to be taken into account in building our future, instead of being ignored in view of adopting models that will be hard to anchor in our realities.

What do you expect from those in charge of official development assistance (ODA)?

Those responsible for ODA are often at the root of our problems. They need to follow us in what we want or just stay home. For us, aid should no longer be considered as a main component. It needs to be refocused on what we are doing, what we have, what we want. It is time to dump the kind of aid that destabilizes, that destroys structures, and acts as lesson-giver. We need to change the paradigm. The options, paths, and directions of those in charge of ODA must not deny our responsibilities for the choices. It is thus more a case of supporting our aims, rather than driving ahead of us in directions that are not of our own making.

At ROPPA, we say that our agricultural policy is our business before being the business of others. We should not be drawing up policies and then waiting for partners to come forward and provide financing. When we develop policies, we should be the first to put up the money to achieve them, meaning that our own countries' budgets should contribute more to their implementation.

Afterward, our partners, in the desire to support us and respect our policy, can come forward to support us financially. This is fundamental, as in our view it is completely out of the question, no matter what kind of aid, regardless of its amount, to force us to deviate from the path that the population as a whole has set out and to define for us the path for emerging from poverty. Unfortunately, we find that every time an African population draws up a policy, the aid system intervenes and asks it to modify its components to the extent that it denatures what we were trying to do. This time, ROPPA is mobilizing in order to communicate our disapproval. We cannot continue, just for the sake of insignificant amounts of aid, to change our policies or deviate from our goals. Today, what we still have to do is to implement this agricultural policy, at the local and regional levels. This assumes that from now on the budgets of our own countries will contribute to assuming responsibility for this agricultural policy.

Positions on Urban-Rural Relationships

Does the urbanization movement currently under way in the West African countries constitute a threat, or instead is it an opportunity for the rural world and agricultural producers?[2]

I do not think that urbanization as such constitutes a threat. With the changes in the economy and society, Africa is called upon to urbanize. Attracted by job possibilities, people will go to the cities. This is the normal course of events. What is troubling about this urbanization is, first, that the cities are built on the wealth produced by the rural world, and there has been no compensation in return. This creates a dysfunctional relationship between the rural world and the urban world. In order to ensure that urbanization is not uncontrolled, an effort should be made to achieve a certain balance between the living conditions of the rural and urban populations. Once this occurs, the rural population will no longer leave the rural world for lack of well-being, simply because there are no possibilities to have a job, but because there is a possibility of a clear improvement in their living conditions. This will leave a certain proportion of the population in the rural world. This is essential, because not everyone can be an urban dweller. There are people who wish to remain in rural areas so long as a minimum number of conditions are met.

Uncontrolled urbanization fails to guarantee any possibility of success. It absorbs resources but offers no advantages. Urbanization is thus a process that is certainly necessary but that needs to be managed in such a way that it is effectively beneficial to all citizens in the region and does not occur to the detriment of the rural world. It thus constitutes a challenge for society as a whole. When cities grow in the absence of any positive outlook for their new citizens, this can be a source of violence. The public authorities have an extremely important role to play in regard to this phenomenon, and they need to regulate while developing urbanization policies and rural development policies that steer clear of this kind of situation. It is up to the public authorities to lay down policies that are distinct from the ones we know today, existing policies that create imbalances between the rural world and the urban world and that give the false impression that a country is no more than its cities. In my own country, for example, you get the impression that Senegal is, well, Dakar! In 2008, the floods in Dakar must have swallowed up vast sums of money, while the groundnut crisis and the locust infestation did not appear to worry anyone. We must not intensify the feeling that rural citizens do not have the same rights as urban citizens. That is the task facing the public authorities.

Is there not a risk that the urban populations will basically consume products from the world market and turn away from local produce?

That is more than a risk. It is a reality. In every city of West Africa people are eating what you can find in Stuttgart, in Paris. This deformation triggered by urbanization is a genuine source of concern; . . . urbanization has not been accompanied by a kind of education between the citizen and the nation. In every country, the citizen needs to have a minimum degree of nationalism. It is in the urban areas that there is the greatest conformity with behaviors from the outside. Under these conditions, the citizens must learn that consuming, for example, groundnut oil produced by Senegalese producers helps the Senegalese economy.

Moreover, access to products from abroad is easier (the products of African farms are considerably less processed). This lack of education, compounded by the ease of access at low prices, prompts people to consume only imported products; that is the reality, and from an economic standpoint it is catastrophic. It changes dietary practices. Ultimately, it will be extremely difficult to reverse course. Today, in Africa's rural areas, people are eating bread, which has replaced our porridge. Even on livestock farms people are consuming powdered milk imported from Europe instead of fresh milk. There is a danger here. The risk is thus here, all too present; it has taken on extraordinary proportions in terms of the deformation of dietary practices, in terms of the consumption of products from abroad—often products that are of mediocre quality. Responsibility for this is shared: regarding that of the citizen, there has to be a little economic nationalism, preference for consuming our own products rather than consuming those from elsewhere; there needs to be work on education, information, heightening awareness. In the primary and secondary schools, there is no type of education demonstrating the nutritional value of niébé, the nutritional value of millet, as compared to corn. Such an educational effort is the responsibility of the public authorities. ROPPA's Afrique Nourricière Program puts forward proposals aimed at solving this problem. In our view, you have to start with young children. The state should show preference for these products as well as create better conditions for bringing them to the cities. It is not acceptable that, throughout the large urban areas, you cannot find any bissap, or any rice from the Senegal River Valley.

Are there any other development paths possible? If so, what are they?

Indeed, there are other paths. This has to be achieved through policies that constitute genuine departures from what is happening now. Principles have to be put on the table and must be accepted.

As far as consumption is concerned, we can move back to policies that show a preference for our national and subregional products. In other regions of the world, such as Europe and the United States, for example, the preference extended to local production through protection has made it possible to meet local demand and then export the surpluses. We are not at the point of negotiating the export of surpluses, we simply want our products to contribute to satisfying local demand. This preferential policy needs to be supported by measures relating to tax policies, information, infrastructure, programs and projects, in sum, by globally cast policies with all sectors. For this to occur there must be participation in policy areas, from their development through implementation, and the stakeholders concerned must be resituated at the core of the solution; unfortunately, the political stakeholders think that they have to do it all.

Be that as it may, for this preference to be effective, it is necessary to ensure that our economies are protected. Indeed, if the current trend with regard to the economic partnership agreements continues, we are headed for disaster. We must have the possibility of protecting our most promising sectors and deregulating others.

The common external tariff of WAEMU is causing us all manner of problems, and there is a risk that the same kinds of problems will arise within ECOWAS. Work can be done on the structures for managing the supply of products such as onions or rice, so that local production can meet the demand for local consumption on a priority basis before outside products are allowed access to the market. Competition with European vegetable oil is also a serious threat for Senegalese groundnuts, affecting nearly 70 percent of Senegalese producers.

As for research, it should focus on the areas that can make it possible to improve the processing of agricultural products. There is considerable work to be done to make up for the current lag. The challenge is to manage to make finished, ready-to-consume products available to the urban populations. This can be done, as it has been in other locations.

Analytical work is also needed to determine the nutritional value of our products, which are often of better quality than imports, and then a public information effort is required so that the citizens are aware of this.

Regarding public services, it is necessary to consider how to improve living standards in rural areas. Currently, access to water and electricity and the availability of leisure activities are still regarded as luxuries, even though, as citizens, rural dwellers must have such access. [...] Once again, it is necessary to promote the availability of these services and the development of small processing units. One possible path could be to reduce the markups levied by the state on the cost of fuel or electricity. These are generally extremely high, and we do not get the sense that the amounts are injected back into the countryside.

Does the countryside have the social and political potential needed to retain its population? Why? How?

Given what is happening, they lack the resources to hold on to their population. Despite the potential, the policies are falling short. Since independence, policies have favored urban development through access to electricity, training, financial resources, sanitation, infrastructure, and so on. The rural areas, referred to as the "backcountry," are trapped in a situation in which they receive very scant resources for their development. National resources are very poorly allocated and very poorly used. The financing of agriculture is not considered as a means of bringing about the development of the land through taking greater advantage of potentials. All the products from the countryside (groundnuts, cotton, coffee, cocoa) are basically processed elsewhere. Moreover, despite the fact that the rural areas make a major contribution to the country's gross domestic product, they do not adequately benefit in return. In fact, not even a part of this wealth is used to promote the creation of added value at the local level, if only through initial primary processing of agricultural products. The most illustrative aspect of this situation is that farmers, after their harvests, follow their products into the capitals only to become "sack toters" and to take on other such menial chores. Thus, in the countryside, the off-seasons are not used to work on processing.

There is no obvious way to retain a rural population—one that is constantly growing—if the pressures on land tenure are not taken into account. It is foreseeable that, in the near future, there will be a need for more aggressive policies on such matters.

This is not to say that a solution is unavailable. The situation can be reversed. There are many possibilities for staunching the bleeding and holding on to a sizable portion of rural dwellers. Industry, based essentially on agrofoodstuff activities, is generally located in urban areas, whereas it could also be situated in the countryside. It is especially at this level that there is potential for basic commodities to be processed, and new services and new trades to be developed—in crafts for example.

This would appear to be all the more essential in that the cities lack the capacity to absorb all those leaving the countryside.

Do African cities have the social and political potential necessary to absorb rural dwellers? Why? How?

Moving into the city is simply the extension of a difficult situation in the countryside. When things do not work out in the city either, efforts are made to go elsewhere. We know that the other African cities have extremely low absorptive capacities. This means moving on to Europe or elsewhere.

If we look closely at the various sectors, we come to the realization that the informal sector is showing signs of weakness, saturation, and a lack of capacity to continue absorbing all those who are coming in from rural areas. This is all the more the case because the development and modernization of trade and industry are such that these sectors require less and less manpower.

Notes

1. Interview conducted by Daouda Diagne, chief of information, communications, and training for Fédération des Organisations un Gouvernementales du Sénégal (FONGS) and a member of the board of Inter-réseaux Développement rural.
2. Interview conducted by Daouda Diagne for the review of the Rural Development Inter-Network (*Grain de sel* 34–35, 2006).

Chapter 12

A Kenyan Agricultural Leader's Standpoint

Philip M. Kiriro*

It is widely recognized today that family farms provide a key to achieving development goals; not only do they withstand adversity, they also organize themselves to diversify and meet the needs of the market in combination with agroenterprises, which must have the necessary infrastructure available. Philip Kiriro calls for greater dialogue between national and regional politicians and farmer organizations; the latter need to make themselves better known and better understood, which will require resources.

What is your vision of the future of family farms in Sub-Saharan Africa and in your region?[1]

In my opinion, African family farms have their whole future lying before them. First of all, the importance of the family farm is recognized by governments and development partners such as the European Commission. If you read the World Bank's *World Development Report 2008,* it clearly emerges that family farms will be the tool that makes it possible to achieve the Millennium Development Goals (MDGs). The European Commission has also very clearly affirmed that achieving the MDGs hinges on the way in which we are going to manage agriculture and, more specifically, family farming. I believe this unanimous recognition is a very positive sign. Threats do exist, of course, but this recognition allows us to believe that family farms are on the right path.

Do you think that small farms are viable from a social, economic, and environmental perspective? How can we avoid increasing the disparities between farmers?

Yes, in my view, the family farm is viable. We need only to make sure that small farming is addressed in an organized way, so that small farmers can produce

*President, Eastern Africa Farmers Federation (EAFF), and Vice-president, Kenya National Federation of Agricultural Producers (KENFAP).

in groups to enable them to approach markets as organized groups, and so that service suppliers can more easily reach small farmers. With organized groups, it is possible to diversify and target needs in order to provide high-value products.

In terms of the environment, the key word is "sustainable" agriculture, which means that we should not focus solely on the economic side of things but also on the environment, and biodiversity. So we are really attentive to these components when we seek to develop the family farm system.

How would you describe the capacity of farmers to implement changes in a rapidly changing world?

Farmers are capable of being agents of change, because we can all benefit from the experience that they have accumulated. Let's not forget that for centuries farmers have fed entire nations, and with very little assistance. Farmers are capable of demonstrating tremendous resolve in the face of adversity. With real organizations, in formalized groups, they can be good partners and at the same time ensure good governance of the agricultural sector.

How would you assess your country's agricultural policies?

It should be said that, in recent times, there have been notable changes. Now, and ever since 2003, policies are negotiated with the farmer organizations. Farmers propose topics of discussion. Governments have even facilitated and rehabilitated marketing initiatives pushed by farmers, such as cooperatives. And farmers have even become involved in the marketing boards that formerly belonged to the government; as a result of the reform process, most are now managed by producers. There are high-quality partnerships at the national level.

Is it possible to develop a common vision between African governments and politicians on the future of the African family farm?

We need to bolster the capacities of politicians. They are, of course, supposed to see themselves as representatives of society, because that is where they come from. But, apparently, many of them do not understand this. Once they are elected, they forget that their strength comes from society itself. We need to help build the capacities of deputies [elected representatives], so they can see the other side of the mirror and be able to work with farmer organizations. In reality, we already work better, in some situations, with governments than with deputies. In Parliament, there are even specialized committees on agriculture that never consult farmers at the national level and never hold a dialogue with them. I think we need to work on that.

What future do you see for youth? Should they stay on the land?

We have a problem to solve before trying to attract young people to farming: we need to make farming attractive. The work of a farmer has to be viewed as a profession that presents challenges and satisfactions, and one that can improve one's standard of living. Today, farming is not attractive, even for parents. That is why many parents try to discourage their children, both girls and boys, from going into farming. So I think we need to institute changes and do some awareness raising. Change needs to come to the agricultural sector if we, as parents, are going to be able to let youth see that a farmer's work is a job with a future.

Can rapid population growth be transformed into a positive fact? Should efforts be made to reduce population growth in the countryside?

Population growth can, by itself, have disastrous consequences. In most countries, population growth is not in sync with growth in food production. We need to manage our population so that the development of agricultural production is adequate.

What place do you see for agroenterprises? What type of relationship with them?

Agribusiness has its place, especially if we have good infrastructure, and if agroprocessing companies do not remain confined to the cities. They should be located around farms, because they then can be a source of dynamics for rural economies. They can also capture some portion of the rural-urban migration and attract youth who are looking for opportunities in the agricultural sector. We need to make sure that the infrastructure exists—power, roads, water supply—so that investors will want to invest in the rural sector.

What is your organization's strategy for promoting the family farm?

In my organization, the East Africa Farmers Federation, we are concerned about the issue of capacity building. We are involved in marketing the output of small farmers, and we hope to use all the advantages of our cooperatives to move into regional markets. So we are fighting for regional integration, for the free movement of persons and goods within the region, and for standardization of our data as a way to facilitate business in the region. We are also concerned about the whole matter of standards—standards that, in our view, should not be aimed only at Europe or European consumers. It is our obligation to have quality standards for our products, so that we can provide our clientele—consumers throughout the subregion—with quality products. Once we know that we are capable of meeting standards and producing unfailing quality, it will be that much easier for us to move into international markets, without having to worry about standards.

What are the problems facing your organization?

Our main difficulty is the challenge of making ourselves known and understood. After all, most of the regional networks are new. In the beginning, our members did not understand what we wanted to do. But now, most of them understand.

Another major challenge is to be capable of playing a role, because the task is huge, and everybody is thinking regionally. The government has changed the policy focus from national to regional. So, as a network, we have a lot of work to do, and our resources are limited.

What do you expect from official development assistance?

Initially, we expect official assistance to be guided by commitments at the national level. I think that governments need to show the way by earmarking greater budget resources for agriculture. In Africa there was the Maputo Declaration, in which governments agreed to increase their budgets.

Development assistance should be used to strengthen government commitments. Another expectation from international assistance is that it should support our strategies aimed at boosting the performance of our farms. It should not be tied to external, predetermined conditionalities. It should instead support our agenda for agriculture.

Note

1. Interview by Anne Perrin, editor, *Grain de sel*, at the Farmers Forum Global Meeting organized by the International Fund for Agricultural Development (IFAD). Original version in English (February 11, 2008) translated into French and subsequently retranslated into English.

Chapter 13

Views of a Burkinabè Small Farmer Leader

François Traoré*

*François Traoré attaches great importance to the value of labor and has faith in what it is possible to achieve.*** *Concerned about the future of young people, he advocates training them and assisting them in becoming involved in family farming. He argues for the strength of agricultural organizations for structuring farmers and serving as a force for proposals to strengthen subsectors and improve agricultural policies. He strongly supports the accountability of all, whether farmers or political leaders.*

A Vision of the Profession of Farmer and the Integration of Youth into Family Farming

François Traoré has been strongly influenced by his life as a farmer and head of household, which he describes as follows:

> I've been a farmer pretty much forever. I became responsible for a family of eight when I was just 16 years old, which certainly taught me very early on how to manage relationships, to earn the trust of the people around me. We were living in Senegal at the time, though we were from Burkina Faso. When I was 20, in 1973, I returned to my village in Burkina Faso, where they raised millet, groundnuts, and fonio. There wasn't any market for any of these crops, and because I wasn't accustomed to farming just for subsistence, I moved to a cotton area in 1979, and then to the village of Sogodjankouli

*Former president of the National Union of Cotton Producers of Burkina Faso, President of the Association of African Cotton Producers.

**The text that follows has been assembled by agreement with François Traoré on the basis of interviews for *Grain de sel* or in conversations with J.-C. Devèze in the 2001–05 period, when Traoré was at Agence française de développement headquarters in Paris participating in a project to support the National Union of Cotton Producers of Burkina Faso, or in other statements.

> in 1980. Starting at about that time, I got involved in association efforts. At first, they were looking for people to handle the provision of inputs as well as marketing within the village-level group. I held the position of general secretary of the group. I became a member of the departmental union of farmer groups in 1991 during a crisis involving the declassification of cotton harvests, which put us at odds with the cotton company. In 1986 I bought my first small tractor. (Interview for *Grain de sel*, 2006.)

François Traoré's thinking is underpinned by a vision of the importance of the family dimension in small farmer agriculture, as shown by the following views expressed during a discussion in 2005:

- "The expression 'involving youth' can be dangerous, as too often it refers to setting them up outside the family, while young people should be assisted by being trained and helped to improve the life of their families. It is a question of 'training genuine leaders who are of value to their families.'"
- "Families split up if power is poorly shared, if the use of the money that is available is not decided upon as a family. . . ."
- "When we talk about gender, we need to include concern about what all members of the family can do: men, women, young people, the elderly. . . ."

Similarly, he attaches great importance to the value of labor, which is particularly valued in the Bwa society, to which he belongs, in western Burkina Faso:

- "You can't just sit under a tree during the dry season; if you aren't working, you are stealing food."
- "My mother reminded me that you get what you want through your own efforts."
- "You have to get beyond a false religious vision by 'forgetting that it is God who does things' and instead have 'a faith that makes mountains.'"

To prepare the future of family farmers, he emphasizes three concerns:

- Having the ability to borrow money for the medium term, with land as security
- Making land tenure more secure, which is important for the sustainability of farms
- Training youth in agriculture and in accountability by helping them ask themselves the right questions that will enable them to commit themselves and make progress

François Traoré, who has three sons of his own, has no problem finding a replacement during his frequent absences from his farm:

> In 1998, at my own expense, I sent my eldest son to Canada for advanced training at a modern farm. Between the three of them, my sons are largely

> responsible for our production. The problem of finding replacements, which is common to the North and the South, is not an issue in my family. After school, my children are comfortable in their fields, and I think they learned a lot with me. The question they ask me nowadays is rather, how we are going to sell all that we produce? (Interview for *Grain de sel*, 2006.)

A Vision for Enhancing the Accountability of Producers and their Organizations

François Traoré works to ensure that producers are at the core of the decisions that relate to them. To do this, he needs organizations whose relative weight makes it possible to take differing views into account. This requires solid structuring of producers (with a single "central unit"), rigorous management of the organizations from bottom to top, and sound cohesiveness between producers and their middle management. This is the context in which it is necessary to consider the participation of producers in the ongoing privatizations in African cotton sectors, taking into account the differing circumstances in various African countries (April 2004 interview). François Traoré argues in favor of providing all farmers with favorable conditions for carrying out their calling, namely, feeding the world (national training workshop of the West African Network of Farmer and Producer Organizations, or ROPPA) and the Small Farmer Confederation of Burkina Faso in July 2005). This is what prompts him to work with the authorities as a "force for proposals" in the definition of agricultural policies.

This is why François Traoré is interested in the conditions that must be met in order to create interprofessional associations:

> The interprofessional association for cotton brings together the three cotton companies and the producers (UNPCB). It is administered by 16 representatives, 8 of whom are cotton producers, and the representatives of the cotton companies. You talk about a phenomenon in vogue when you refer to the interprofessional associations; for us, it is a genuine obligation. We have no special relationship with the other interprofessional associations, but if you were to ask us about the grain interprofessional association, I would have to tell you it leaves me a bit dubious. That structure was created without organizing and adequately structuring its members, be it the producers or the other families. An interprofessional association has to be able to intervene to boost the value of agricultural products, but in the case of cereal grains, for the time being there is no way to know the amount of production or to determine how much is available for marketing. As to the role of the state, it seems normal to me that it encourages this kind of initiative, but not without structuring the stakeholders. It is up to the producers in particular to see to this; however, the Burkina Faso Small Farmer Confederation was unable to help out with structuring the cereal grains producers, in contrast to the situation with rice, where it was possible to assist with the creation of a national structure for rice producers.

> Moreover, producer pressure is beginning to make itself felt in the activities of that interprofessional association." (Interview for *Grain de sel*, 2006.)

François Traoré would give the following advice to a young leader of an African producer organization: "Endurance, desire, and a spirit of cooperation are, to my way of thinking, extremely important. Everyone is at risk on earth—the cotton producers just like everyone else. You have to believe in order to genuinely exist and dare." (Interview for *Grain de sel*, 2006).

A Vision for the Organization of Cotton Producers

Here is how François Traoré presents the birth of the Association of African Cotton Producers (APROCA), which he chairs:

> In 2001, I was alarmed by the drop in cotton prices. Between January and October of that year, the world market price for cotton collapsed. It lost half its value. We tried at the time to understand why the price drop had occurred. After some investigation, we came to the realization that the subsidies were the cause. On November 21, 2001, we published an article in the UNPCB newspaper, the quarterly entitled Le producteur, in which we denounced these subsidies and included a letter to that effect. In the course of one of his visits, Father Maurice Oudet proposed that we publish our letter on its listserv. The organizations of West African cotton producers that read it on the site expressed their support for us. Benin and Mali responded quickly. For that reason, in January 2002 we launched our second appeal, the joint appeal of West African cotton producers.
>
> That was the start of the cotton mobilization, which would result in the mobilization we know of on the international level, in which the producers are playing a role at the top level alongside their governments. While today the subject is blessed with broad media coverage, that wasn't the case at the time. At the meeting of the WTO[World Trade Organization] in Cancun, I somehow acquired the image of a representative of cotton producers. But I was in a weak position, as I was all by myself. Subsequently, the cotton producers from a number of countries asked: "Why don't you bring us together yourself?" So then we organized a meeting mobilizing the cotton producers from several African countries, where each participant assumed the costs of his own travel. That meeting was held in Cotonou, in December 2004, with representatives from six countries (Benin, Burkina Faso, Cameroon, Mali, Senegal, and Togo), and APROCA was born.
>
> Today, APROCA brings together producer representatives from 13 African countries (Benin, Burkina Faso, Cameroon, the Central African Republic, Chad, Côte d'Ivoire, Gambia, Ghana, Guinea, Guinea-Bissau, Mali, Senegal, and Togo). At the World Cotton Organization meeting in Hong Kong, APROCA sent 18 representatives. Our permanent secretary, a national of Côte

> d'Ivoire, is based in Bamako. His name is Mamadou Ouattara. We have just recruited a Malian accounting officer. APROCA receives subsidies from various donors (Oxfam, ENDA, French Cooperation, ICCO, SNV). In 2006, we hired a Burkinabè spokesperson and are now working to communicate in all languages of the member countries (French, English, and Portuguese); international negotiations are often conducted in English.
>
> With regard to relations with ROPPA, I should state at the outset that I myself am a ROPPA Board member. ROPPA had organized a workshop on the Cotton Observatory, in the course of which the producers indicated that they wanted a specific structure. Nothing moved, but it was an urgent matter. I told them about our initiative, and ROPPA did not want to get further involved, arguing that it wasn't in a position to set up branch offices everywhere. Although ROPPA doesn't support us in this initiative, we have made progress, which is what the producers asked for.
>
> In the various countries, privatizations are under way. In Africa, there is an association for cotton ginners. When up against professionals, it is essential that the producers be represented by specialized professional organizations that are stronger, failing which we will be at a disadvantage in negotiations. This said, ROPPA still presents itself as the representative of African cotton producers in Brussels. (Interview for *Grain de sel, 2006.)*

This is how François Traoré presents his vision of the role of APROCA in the same interview:

> APROCA has never had the unique role and objective of combating the subsidization of cotton in the countries of the North. Our struggle, as noted in our goals—which I invite you to consult on our website at www.aproca.net—is in several directions:
>
> - increase the degree to which producers are structured;
> - support the involvement of producers in the various subsectors; and
> - promote exchanges of producers among the various structures and various countries.
>
> Combating subsidies is, of course, crucial, but this struggle at the international level is not the essential point for us. As to the possible decline in competitiveness of African cotton, it is certain that one has to be prepared for any eventuality. But this is the case for everyone in every area who seeks to achieve a better level and remain there. Productivity gains for cotton can be found. I made reference a moment ago to the strategies introduced by farmers to address the lack of inputs. Increasing soil fertility, improved technical management of crops, and better management of income from the farm are among the solutions that need to be pursued in order to achieve productivity gains.

A Vision of the Future of African Agriculture and Development of the Continent

An open letter addressed to Tony Blair in 2005, when François Traoré was chairman of the Burkina Faso Farmers Confederation (CPF)—long excerpts of which are provided below—gathers together the key points of his vision as a leader devoted to the development of Africa. [Text from the official English at http://www.abcburkina.net/content/view/140/45/lang,en/]

1. We, too, are convinced that *Africa is not condemned to remain in poverty and destitution.* A few apposite measures at the international level and some others at the regional and national levels would be enough.

2. *Europe has no advantage in seeing Africa sink into underdevelopment and destitution.* Europe developed itself industrially and agriculturally behind protection barriers. The priority is not Africa's integration into the world market, but the acceleration of its development at its own pace and through its own channels, especially not to reach deadlock on agriculture!

3. *No development will be possible as long as the majority of the population—consisting of farmers and breeders—is not taken into account.* In addition, the majority of poor people on this continent are in the rural areas. In Burkina Faso, for instance, 92 percent of poor people live in rural areas. That is why African countries asked that agriculture be on the agenda for negotiations that took place between the European Union and the APC (African, Pacific and Caribbean) countries in the context of the Cotonou Accords. Europe, which talks about partnership, refused. This decision has to be reversed. All the "partners" must commit themselves to refusing to sign the APE (Accord for Economic Partnership that Europe wants the APC to sign) without those different regions of the APC countries defining a worthwhile Common Agricultural Policy (CAP).

4. Moreover, *how can you imagine helping Africa by setting up free trade between farming produce from Europe (which benefits from high technology and heavy subsidies) and that of Africa, whose producers are poorly equipped in farming materials and are not subsidized?* The CAP (Common Agricultural Policy) must therefore be consolidated by a rate of protection sufficient to keep farmers safe from low-priced imports (by-products and others) and uncontrolled variations on the world market (a drop in the dollar). Most effective would be to withdraw a certain number of sensitive products (rice, milk, wheat, jam) from the "CET" (Common Exterior Tariffs), for which import taxes would be defined, leveling variable rates in function of the world market, thus allowing guarantees of a stable income for farm producers as well as breeders.

5. It is essential also that *countries wishing to help us with food aid not take advantage of it to help their own farmers.* If they really wish to help us, let them not just dump their old stock, but give financial aid instead. With that

money, our countries will be able to buy food supplies from surplus zones (there is no lack of them!) and transport them to zones where food security is temporarily under threat. In this way, populations in difficulty will be helped and producers who have had a good harvest will be able to sell their produce. This would be a great step forward, as today African farmers are the biggest losers in the present system. When there is not enough rain, they do not have a good crop, and when the rain allows them to have a good harvest, they cannot sell their produce!

6. Europe is preparing to defend its new Common Agricultural Policy (CAP) at the next ministerial conference of the WTO to be held in Hong Kong in December 2005. *We too will then have our CAP to defend!* (CEDEAO Agricultural Policy). We count on Mr. Blair, that Europe will help us defend our CAP, as well as its cotton dimension.

7. Finally, in spite of some debt cancellations, most African countries are collapsing under the weight of the debt. We have seen how generous Mr. Blair was towards Iraq and the Asian countries. Must there be other disasters this time in Africa so that African countries will have their debts cancelled? Why not let agriculture benefit from the new debt cancellations that you will certainly succeed in acquiring? *As agriculture is the engine of development, it would be sensible if the HIPC [Heavily Indebted Poor Country] initiative were not restricted only to social sectors such as education and health.*

François Traoré revisited the subject of the responsibility of political leaders in an interview with Falila Gbadamassi for www.Afrik.com on February 22, 2007:

> The African Heads of State should continually remind their counterparts in the industrial countries that they are not following the rules that they themselves instituted, and that they are working against poverty reduction even as they say it is a priority. None of the rich countries will slide back into poverty if they endeavor to make this message a reality, because they have, simply put, more opportunities to adapt than the poor countries do. Nothing is keeping them from continuing to manufacture ever more sophisticated weapons, even if no one knows who they are going to destroy with them. All these large countries have nuclear weapons, even though you can only destroy the Earth once. Just because you are rich doesn't mean you can't be mistaken, and I think that the major economies have erred in not putting the economy to work to serve mankind, but instead to serve only a limited number of individuals.

Chapter 14

Rethinking Agricultural Policy

Ibrahim Assane Mayaki*

Caught between poverty reduction and support for the private sector, public policy struggles to take into account the family farm, even though the family farm is at the center of the challenges facing Sub-Saharan agriculture. It is only through dialogue among the parties concerned that it will be possible to rethink agricultural policy. This will require articulating national and regional public policies in the context of international negotiations that give voice to the silent majorities.

An analysis of policy making by government entities in Sub-Saharan Africa must look at the characteristics of these governments and their current modes of operation. Too often the emphasis is placed on an analysis of the (so-called technical) instruments contained in their policies and the "rationality" of these instruments, with scant regard for the interests and power relations of the various parties involved in their definition and implementation. Inevitably, as a result, the conclusions generally drawn from these analyses repeatedly come up against the same obstacles and churn out the same solutions.

Policy Blindness in the Face of the Evidence

It is genuinely difficult for agricultural policies to take full account of some of the evidence and the main current challenges. First of all, the vast majority of farms in Sub-Saharan Africa are family farms. Second, while there is substantial urban drift in most Sub-Saharan African countries, the rural population will continue to grow, as will the number of family farms (figure 14.1). Finally, family farms will have to take up the agricultural challenges facing Sub-Saharan African countries over the next 20 years. These challenges involve the productivity of the agricultural sector (land, labor, and capital), land use planning and the integrated management of natural resources, the integration of young people in rural areas, and access to markets.

*Former Prime Minister of Niger, Former Executive Director of the Rural Hub, Dakar; Chief Executive Officer of the New Partnership for Africa's Development (NEPAD) since January 2009.

Figure 14.1 Rural, Agricultural, and Urban Population Trends in West Africa

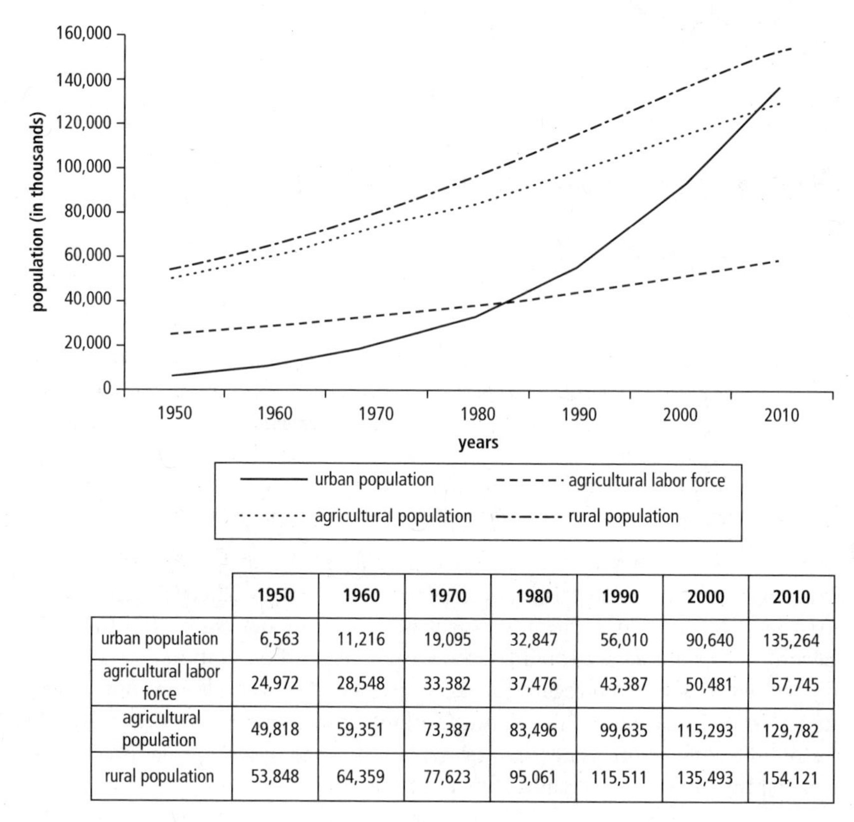

	1950	1960	1970	1980	1990	2000	2010
urban population	6,563	11,216	19,095	32,847	56,010	90,640	135,264
agricultural labor force	24,972	28,548	33,382	37,476	43,387	50,481	57,745
agricultural population	49,818	59,351	73,387	83,496	99,635	115,293	129,782
rural population	53,848	64,359	77,623	95,061	115,511	135,493	154,121

Source: HUB RURAL, Dakar.

Differentiation of Farms and Public Policies

Clear distinctions began to be seen in the agricultural sector in Africa in the early 1990s. Some authors have clearly depicted the processes under way, identifying three types of agriculture:[1]

- big-business agriculture, emerging from the best-equipped family farms or having benefited from direct investment in the most promising subsectors;
- a growing fringe of marginal farms that no longer have the resources to ensure their survival and that find themselves in a downward slide into poverty; and
- a large number of farms between the two extremes, on the razor's edge as a result of market instabilities or natural disasters.

Public policy accentuates this phenomenon by focusing successively on often conflicting goals: on the one hand, poverty reduction based on social action in favor of the poor; on the other, support for the private sector, which emphasizes the competitiveness of agroexporting enterprises. This leads to acrimonious confrontations between the public authorities and farmers' movements regarding the reform of agricultural policy instruments: technical services, farm credit, and vocational training for farmers.

Reforging public policy in the field of agriculture by focusing it on genuine support for farmers is essential for our agricultural development.

Investing in Human Capital

The agricultural sector has gradually been depleted of its resources (particularly its human resources) since the late 1980s. Whether in terms of the level of technical training of farmers or the quality of supervision and of senior officials, a general erosion of human capital has been observed and, ultimately, has severely limited the capacity of the sector to attract investment (including from the government budget) for its economic development. Although statistics show the relative importance that this sector still has in the economies of most of the Sub-Saharan African countries, it remains marginalized and receives very few fiscal resources. Why is this the case? The answer lies both in the difficulties that experts in the sector face in demonstrating the relevance and effectiveness of such investment and in the public decision makers' lack of interest in agriculture. Current poverty reduction strategies will not be sufficient to change this tendency.

Professional agricultural organizations have a key role to play, along with the authorities of their countries and the two regions (West Africa and Central Africa), in strongly advocating investment in the agricultural sector by putting forward different development models. In this context, it will be essential to shift the current paradigms by focusing more closely on the issue of reducing inequalities, which must be analyzed in depth to construct these alternative development strategies.

Constructing a Frank and Serious Dialogue on Policy

The quality of the processes leading to the adoption and implementation of public policies is key. These include not only upstream analyses and discussions among the parties concerned, but also appropriate decision-making mechanisms. There is frequently a mismatch between decisions actually made and the strategic options that were supposed to underlie these decisions. Political

Figure 14.2 Consistency in the Dialogue

stages	consistency	inconsistency	role of civil society
vision/social demands			+
sectoral assessments			+/–
formulation of agricultural policies and strategies			–
implementation of strategies			–
ex post assessments			+

Source: HUB RURAL, Dakar.

trade-offs are made and are another area in which action is essential. A policy is necessarily the result of a power struggle between social and other interest groups and the public authorities. As a result, simply working on the tools is not enough.

As figure 14.2 shows, the dialogue with the parties concerned (particularly civil society) is cultivated during the construction of the "vision" and in the "ex post assessment" but is entirely neglected during policy making and implementation.

Articulating National and Regional Public Policies

Designed and implemented by the public authorities in the aftermath of periods of social, political, and especially economic instability in Africa in the 1980s and 1990s, integration policies in general, and regional agricultural policies in particular, have almost the same goals as national policies. Bound by one and the same context—that of withdrawal of the state from the agricultural sector—they lead countries to the same finding: a recognition of the essential role that the agricultural sector has in the economy. They are also bound by identical aims, such as achieving food security based on an adequate level of self-sufficiency in their territorial jurisdictions. However, beyond these convergences, the policies are quite distinct in terms of the processes and instruments used to achieve their goal. This difference between the regional and national levels is largely influenced by one of the key principles—subsidiarity—which underlies the complementarity between the two levels.

While regional agricultural policies endeavor, by promoting intraregional trade, to enlarge the regional market and find outlets for community products, national agricultural policies place greater emphasis on increasing productivity and output. This explains the complementarity of these two types of agricultural policy. At the national level, markets are too small to provide solutions to all the problems of food insecurity, while at the regional level there is insufficient infrastructure to meet the production challenges and respond to the needs of consumers and markets. Regional agricultural policies therefore cannot replace national agricultural policies, which are essential, but which have been stripped of their main support instruments.

Articulating Regional Policies in International Negotiations

The main challenges in the current trade negotiations relate to the need to reforge agricultural policies with a focus on support for food products. The food deficit (in volume terms) of the Economic Community of West African States (ECOWAS) increased threefold between 1995 and 2003. And yet the ECOWAS countries base their economic policies on the growth of the agricultural sector. To ensure the effectiveness of these policies, the linkage with the trade negotiations currently under way is clear.

The tight timetables for the negotiations leave African countries little room to maneuver in a context of overlapping agendas for the Millennium Development Goals and the EPA (economic partnership agreement between the ACP regions and the European Union). The main issues to be dealt with are special products and appropriate safeguard mechanisms, but the West African and Central African countries have not yet made any "real" proposals on these issues. This is paradoxical in that one of the critical points in the EPA process is the definition of sensitive products to be excluded from the liberalization of trade with Europe.

Although significant progress has been made, the conditions are not yet present for a dialogue between the various parties concerned, which limits the "normal" progress of the negotiations. In the case of economic partnership agreements, it is essential to agree on a consistent definition of a program for upgrading the productive sectors. This program will constitute the foundation for the rural and agricultural development strategies. If the removal of trade barriers is not accompanied by specific upgrading measures, the economies of the region will be seriously weakened.

For the agricultural sector, it will therefore be essential to assess the adjustment costs so that the necessary measures may be defined and financing needs accurately identified.

Constructing and maintaining a solid basis for dialogue involving all the parties concerned—including, in particular, civil society—to ensure progress in the negotiations and thus to maximize the chances that West and Central Africa will have "intelligent" room to maneuver is essential to achieving the objectives of the agricultural policies.

Today, 60 percent of Sub-Saharan Africa's assets are found in the agricultural sector, but their influence in the formulation of public policy affecting the sector is very limited; remedying this will certainly increase the effectiveness of our policy making and above all, our democratic governance. We have 50 years of experience of what not to do.

Wittgenstein said that science helps to solve all the unimportant problems. I understand that to mean that beyond the sophistication of the (necessary) planning tools, it is essential to achieve a new political will that gives voice to the silent majorities. It is only in this way that Africa can really move forward.

Note

1. J.-F. Bélières, B. Losch, and P.-M. Bosc. 2002. "What Future for West Africa's Family Farms in a World Market Economy?" Issue Paper E113. International Institute for Environment and Development, London (October).

Chapter 15

In Conclusion, a Question: Can Tropical Africa Be a Future Agricultural Giant?*

Jean-Claude Devèze

In response to the opinions voiced about agriculture in Sub-Saharan Africa, which are so often clouded by conventional wisdom or beset with doubts and pessimism, this book has three goals. The first is to provide readers with data and ideas that are sufficiently developed to enable them to forge their own opinion on the nature of the agricultural challenges facing Africa and on what can and should be done to meet these challenges. The second goal is to clearly explain what is at stake in terms of the future of the African family farm and the importance of taking better account of the positions of the agricultural leaders who are working to promote and develop such farms. The third is to shed light on the many changes under way, which prove that this mode of farming is anything but static and mired in the same old unimaginative solutions; it can indeed play an increasingly important role not only in providing food for the African people but also in contributing to the development of the subcontinent and its better integration into global trade.

The current state of African agriculture and its low productivity can lead, not unnaturally, to a pessimistic view of its ability to close the gap with other forms of agriculture around the world. Although one should be wary of mechanically copying methods and techniques from one continent to another, rural histories in other parts of the world do show that agricultural development can take many forms. Asian agriculture—based on irrigated rice cultivation and associated with ancient civilizations, solidly structured societies, and relatively effective state entities—has been able to adopt institutional and technical processes suited to production and trade that have enabled it to rapidly increase yields and cope with population growth and urbanization. Agriculture in the

*The title of French geographer Pierre Gourou's most recent book is: *L'Afrique tropicale, nain ou géant agricole?* (Flammarion, 1991). A previous book was *Terres de bonne espérance: le monde tropical* (Plon, *Terre humaine* Collection, 1982).

developed countries has also increased yields as well as lands under cultivation owing to the combined effect of rural-urban migration, which has led to the sustained development of other sectors of the economy, and large investments over the years, including investments in training.

African agriculture must find its own way to meet the many challenges facing it, including not only feeding the rapidly growing population of the subcontinent but also providing jobs and income for the cohorts of young people arriving on the job market, promoting the development of other sectors, acquiring foreign exchange, strengthening territorial dynamics, and controlling energy consumption, among other things. African agriculture has shown its capacity to adapt to diverse and changing natural and economic conditions by combining recourse to endogenous innovations and local savoir faire with prudent incorporation of exogenous ideas that often involve new risks. In many cases, cross-fertilization between the endogenous and the exogenous leads to "virtuous" processes, such as mixed farming, agroecology, the selection of priority investments, the introduction of new institutional approaches (such as farmers organizations), the establishment of interprofessional umbrella organizations in the subsectors (value chains), and the introduction of farm management advisory boards.

The reflections of the various contributors to this book have most of all shed light on the importance of placing greater emphasis on the human and natural potential to meet these challenges, at the risk of seeing an upsurge in humanitarian crises and conflicts and watching an entire continent drift away from the rest of the planet.

The time when low-price agricultural surpluses made it possible to meet the needs of African urban populations seems to have passed, since agriculture in the industrial countries—mechanized and dependent on chemical inputs—will have difficulty continuing to increase its yields and since land reserves are limited, except in Africa and Latin America. So, despite sometimes difficult climatic conditions and land that is often lacking in fertility, it will be necessary to mobilize the natural potential of the tropics while taking account of their special characteristics. In terms of additional biomass output, their potential is substantial, as long as available land, water, and genetic capital are better managed.

This will require the mobilization of human resources, starting with the holders of family farms, who make up the vast majority of farmers and who will continue to do so for decades to come. It will be necessary to help them to manage change, both on their farms and in their communities, in order to increase productivity. This will require training men and women and organizing them professionally and collectively to enable them to make their voices heard and defend their interests against other economic operators—agroindustries, processing companies, merchants, and, in particular, the state.

Two complementary rural and agricultural development approaches are proposed for the mobilization of human capital and natural potential, one based on strengthening local dynamics to develop the production areas and territories in which family farms are located, and the other favoring increased consideration of agriculture at the national and regional levels in the context of development of the rural economy and its links with other sectors and, more generally, with the global markets. However, public policy will have to provide rural areas with essential services and infrastructure and help to build a favorable economic and institutional environment that inspires confidence. It will be necessary to promote the transition from agricultural units that, of necessity, are dedicated to the survival of the family to other modes of farming adapted to the characteristics of the ecosystems, the commercially most advantageous products, the markets targeted, the potential capital intensiveness, and the necessary degree of technical capacity. At the same time, the following two cautions are necessary: not to destroy—or, failing that, to reconstitute—the natural capital on which agricultural activities are based to ensure their sustainability; and to provide strong support for alternative activities for those who leave the land, particularly the many young people, at the risk of marginalizing them and endangering the social and political peace.

While it seems possible from a technical, economic, and social standpoint to meet the various agricultural challenges, the primary difficulty is political. A conviction shared by most of the authors of this book, and in particular by agricultural leaders, is that African policy makers have until now not had the political will to cooperate with the agricultural profession in promoting this sector. They do, of course, have their excuses: the African continent is recently independent, balkanized into economically modest countries, entering globalization at a poor level of competitiveness, endowed with government entities that are often lacking in resources and mired in the structural adjustments of previous periods, and confronted with the difficulty of mustering its forces in the emerging regional forums. However, caught between the private interests of the very wealthy and the need to feed largely poor urban populations, the question is whether policy makers have finally understood the importance of making the agricultural sector a priority for the long term, which will require, in particular, guaranteeing prices that are high enough and stable enough to mobilize farmers whose only wish is to farm.

The recent return to favor of agricultural development as a key tool of poverty reduction and as a major factor alongside demographic policies may signal a sea change, both among African leaders and among the most important providers of development assistance. This book will have achieved its purpose if it helps to better anchor this change at a time when a new order can and must allow farmers in Africa to take their rightful place, as they do in Brazil, which has lifted itself to the rank of agricultural giant in the space of two decades.

Index

Boxes, figures, notes, and tables are indicated by b, f, n, and t following page numbers.

M

ECO-AUDIT

Environmental Benefits Statement

The World Bank is committed to preserving endangered forests and natural resources. The Office of the Publisher has chosen to print ***Challenges for African Agriculture*** on recycled paper with 50 percent post-consumer waste, in accordance with the recommended standards for paper usage set by the Green Press Initiative, a nonprofit program supporting publishers in using fiber that is not sourced from endangered forests. For more information, visit www.greenpressinitiative.org.

Saved:
- 9 trees
- 3 million British thermal units of total energy
- 849 pounds of net greenhouse gases (CO_2 equivalent)
- 4,087 gallons of waste water
- 248 pounds of solid waste